Springer Theses

Recognizing Outstanding Ph.D. Research

For further volumes:
http://www.springer.com/series/8790

Aims and Scope

The series "Springer Theses" brings together a selection of the very best Ph.D. theses from around the world and across the physical sciences. Nominated and endorsed by two recognized specialists, each published volume has been selected for its scientific excellence and the high impact of its contents for the pertinent field of research. For greater accessibility to non-specialists, the published versions include an extended introduction, as well as a foreword by the student's supervisor explaining the special relevance of the work for the field. As a whole, the series will provide a valuable resource both for newcomers to the research fields described, and for other scientists seeking detailed background information on special questions. Finally, it provides an accredited documentation of the valuable contributions made by today's younger generation of scientists.

Theses are accepted into the series by invited nomination only and must fulfill all of the following criteria

- They must be written in good English.
- The topic of should fall within the confines of Chemistry, Physics and related interdisciplinary fields such as Materials, Nanoscience, Chemical Engineering, Complex Systems and Biophysics.
- The work reported in the thesis must represent a significant scientific advance.
- If the thesis includes previously published material, permission to reproduce this must be gained from the respective copyright holder.
- They must have been examined and passed during the 12 months prior to nomination.
- Each thesis should include a foreword by the supervisor outlining the significance of its content.
- The theses should have a clearly defined structure including an introduction accessible to scientists not expert in that particular field.

Thomas Sokollik

Investigations of Field Dynamics in Laser Plasmas with Proton Imaging

Doctoral Thesis accepted by
Technical University, Berlin, Germany

 Springer

Author
Dr. Thomas Sokollik
Lawrence Berkeley National Laboratory
Mail Stop 71R0259, 1 Cyclotron Road
Berkeley, CA 94720
USA
e-mail: TSokollik@lbl.gov

Supervisor
Prof. Wolfgang Sandner
Max-Born-Institute
Max-Born-Str. 2a
12489 Berlin
Germany
e-mail: sandner@mbi-berlin.de

ISSN 2190-5053
e-ISSN 2190-5061

ISBN 978-3-642-26677-5
ISBN 978-3-642-15040-1 (eBook)

DOI 10.1007/978-3-642-15040-1

Springer Heidelberg Dordrecht London New York

Cover design: eStudio Calamar, Berlin/Figueres

Printed on acid-free paper

Springer is part of Springer Science+Business Media (www.springer.com)

Supervisor's Foreword

Laser plasma physics was established almost immediately after the invention of the laser, now exactly fifty years ago. Pulsed laser light, even from early lasers, can easily be focused to sufficiently high intensities for multi-photon or field ionization to occur, turning matter into a plasma state. In addition, the momentum transfer from the light pulse, directly proportional to the intensity, can be used to compress the plasma, which can be further heated by a variety of light-matter interaction phenomena. Hence, one may expect the plasma density and temperature to depend crucially on the intensity of the focused laser light.

Laser technology has progressed tremendously over the last fifty years in producing ever higher light intensities. One of the crucial parameters is the pulse duration, which has advanced from nano- to pico- to femto-seconds. A tremendous breakthrough was the implementation of the chirped pulse amplification technique in the mid 1980s. It allowed the amplification of very energetic short pulses without self-destruction of the laser and finally led to the creation of light fields of so-called relativistic intensity. Electrons in such fields are accelerated close to the velocity of light and their movement is governed by the laws of relativistic kinematics. The properties of the laser—plasma interaction are influenced significantly by these effects. With modern lasers, reaching intensities well in excess of 10^{20} W/cm^2, the relativistic region is easily reached and extremely high plasma energy densities can be achieved.

One of the consequences—potentially of considerable importance to society— is the fact that the laser plasma is a brilliant source of photon and particle radiation. While electro-magnetic radiation is expected under such conditions, laser plasmas as a source of particle beams are less obvious and actually represent a relatively young research field. It was discovered only about ten years ago that, in the interaction of relativistic laser intensities with solid target foils, a considerable part of the incident laser energy can be converted into kinetic energies of fast ions. These arise either from contamination layers of the foil or from the target substrate itself. This observation has triggered a wealth of theoretical and experimental investigations. By now a variety of acceleration mechanisms has been identified, including direct acceleration of ultra-thin foils through momentum transfer from the light pulse (radiation-pressure acceleration).

From an applications point of view (and compared to conventional accelerators) laser acceleration of ions is still at its infancy. Parameters like monochromaticity, beam stability and average power still need considerable improvement in order to become competitive. One fascinating property of laser-accelerated ions, however, is the fact that ion trajectories in the emitted beam show a laminar behavior with nearly no crossing. The measure of this quality—the emittance—is two orders of magnitude better than in conventional ion accelerators.

Such emittance allows for excellent resolution in imaging applications like proton radiography. This is where the thesis of Thomas Sokollik takes up the challenge. Specifically, he has developed a novel imaging technique and is the first to apply it to measuring both the spatial and temporal evolution of ultrastrong electrical fields in laser-driven plasmas. Proton radiography or simply proton imaging of laser produced plasmas is performed here with two intense laser pulses. One pulse creates the plasma which is to be probed, and the other produces a proton beam acting as the probe. Hence, the laser-created protons helped in imaging and understanding their own creation process.

Under the laser parameters of this work the Target Normal Sheath Acceleration (TNSA) has been identified as the dominating ion acceleration scheme. Thereby electric fields up to the order of MV/µm, arising from charge separation in the laser-created plasma, accelerate ions to kinetic energies in the MeV energy range, depending on the irradiation conditions. The fields arise on a pico-second time scale, posing considerable challenges to any real-time probing scheme.

The advantage of laser driven proton/ion beams for imaging purposes are their short emission time (of picosecond duration) and the low transversal emittance. In addition, these beams usually have a broad kinetic energy distribution which is considered a disadvantage for many applications, but has been turned into an advantage in the present set-up. The introduction of a drift length between proton creation and plasma sample results in a temporal separation of protons from different velocity classes and, hence, in a temporal resolution of the probing process. The temporal resolution is preserved if the kinetic energy of the probing protons is recorded together with their deflection due to the sample fields. Most elegantly, this can be done in a dispersive spectrometer, where the proton velocity (corresponding to the probing time) is expanded along one axis while the field-induced proton deflection occurs on an orthogonal axis. This method has been developed in Thomas Sokollikt's PhD work and has been called "proton streak deflectometry". The method allows the continuous recording of the temporal evolution of a strong field within a thin, two-dimensional probe layer.

The dynamics of electric fields which drive the ion acceleration depend crucially on electron transport processes in different target systems. Using the proton streak deflectometry it was possible to show that lateral electron transport is one of the key processes which determines extended field distributions and geometries that influence beams of charged particles emerging from extended foil targets. Consequently, the field dynamics should change if one can impose constraints on the lateral electron transport at the target system. Therefore, investigations were extended to spatially confined targets, particularly micro-spheres or -droplets which were earlier successfully used to create quasi mono-energetic proton beams. The confined geometry of plasmas and fields was found to have positive effects on the kinetic energy and spatial distribution of accelerated ions. This was shown both in experimental radiography images and in numerical simulations, one of which was selected for the cover page of Physical Review Letters. Finally, the results of the thesis have triggered a very sophisticated new interaction experiment with single levitated microspheres in a field trap. These recent results have shed light on a surface plasma effect which had not been considered for isolatedmicro-targets so far.

Time resolved proton radiography, as first applied in the present thesis, is not only among the very first scientific applications of laser-accelerated protons. More generally, the interplay between comprehensive diagnostics and full experimental control over laser plasmas is the key to future optimization of laser-driven particle acceleration itself. In the present work the utilization of the intrinsic low particle beam emittance has already led to new insights into the plasma field dynamics which, in turn, influence the energy distribution of laser accelerated protons. If progress continues along such lines one may expect the vision of a new era of accelerator physics and novel applications to come true.

Berlin, December 2010   Wolfgang Sandner

Acknowledgments

I would like to thank all those people, who have contributed to this work. Without their help, constructive discussions and friendship this theses would not have been possible.

Special thanks are given to the following people:

- Prof. Dr. W. Sandner for the possibility to work in this excellent research group and his commitment in the Ph.D. seminar.
- Dr. P.V. Nickles for his dedication to the whole department.
- Dr. M. Schnürer for his outstanding competence, his support and mentoring.
- Dr. G. Priebe for much more than keeping the glass laser running.
- Dr. Ter-Avetisian for fruitful discussions.
- Dr. H. Stiel and Dr. H. Legall for reviewing an early version of this manuscript.
- Prof. Dr. A.A. Andreev for answering all my questions concerning theoretical issues.
- S. Steinke for friendly support and visionary discussions.
- P. Friedrich for extraordinary commitment and support.

and all other administrative and technical employers: S. Szlapka, B. Haase, G. Kommol, J. Meißner, J. Gläsel and D. Rohloff.

This work was partly supported by DFG—Sonderforschungsbereich Transregio TR18.

Berlin, December 2010   Thomas Sokollik

Contents

1 Introduction 1
References 3

Part I Basics

2 Ultra Short and Intense Laser Pulses 7
2.1 Mathematical Description 7
2.2 Single Electron Interaction 10
2.3 Ponderomotive Force 12
References 14

3 Plasma Physics 17
3.1 Light Propagation in Plasmas 17
3.2 Debye Length 20
3.3 Plasma Expansion 21
References 23

4 Ion Acceleration 25
4.1 Absorption Mechanisms 25
4.1.1 Resonance Absorption 25
4.1.2 Brunel Absorption (Vacuum Heating) 27
4.1.3 Ponderomotive Acceleration, Hole Boring
and $j \times$ B Heating 28
4.2 Target Normal Sheath Acceleration 29
4.3 Alternative Acceleration Mechanisms 32
References 34

5 Laser System 37
5.1 Ti:Sa Laser System 37
5.2 Nd:glass Laser System 40

5.3   Synchronization   41
References   44

Part II   Proton Beam Characterization

6   Proton and Ion Spectra   47
6.1   Thomson Spectrometer   47
6.2   Quasi-Monoenergetic Deuteron Bursts   50
6.3   Irregularities of the Thomson Parabolas.   51
References   52

7   Beam Emittance   55
7.1   Virtual Source   56
7.2   Measurement of the Beam Emittance   57
References   59

8   Virtual Source Dynamics   61
8.1   Energy Dependent Measurement of Pinhole Projections   61
8.2   Shape of the Proton Beam.   64
8.3   Energy Dependence of the Virtual Source   67
References   67

Part III   Proton Imaging

9   Principle of Proton Imaging   71
9.1   Principle Experimental Setup.   71
9.2   Gated Multi-Channel Plates   73
9.3   Time Resolution.   74
References   75

10   Imaging Plasmas of Irradiated Foils   77
10.1   Experimental Setup.   77
10.2   2D-Proton Images.   78
References   81

11   Mass-Limited Targets   83
11.1   Experimental Setup.   84
11.2   Water Droplet Generation   85
11.3   Proton Images of Irradiated Water Droplets.   87
11.4   3D-Particle Tracing.   91
References   93

12 Streak Deflectometry 97
  12.1 The Proton Streak Camera. 97
  12.2 Streaking Transient Electric Fields 99
  12.3 Fitting Calculations. 101
  12.4 Particle Tracing 103
  References 106

13 Summary and Outlook 107

Part IV Appendix

14 Zernike Polynomials 111
  Reference 112

15 Gated MCPs 113

Curriculum Vitae 117

Index 121

Chapter 1
Introduction

Since the invention of the laser in the year 1960, a continuous progress in the development of lasers has been made. Especially with the "Chirped Pulse Amplification" (CPA) technique invented in 1985, a rapid enhancement of the laser intensity was achieved in the last two decades which is still going on. The pulse duration has been decreased down to a few femtoseconds. By focusing these pulses tightly to several micrometers in diameter huge intensities are reached. The interaction of these intense and short laser pulses with matter causes multifarious phenomena which are in the focus of recent investigations. In general, one could say that the laser pulse ionizes the atoms, creating a mixture of free electrons and positively charged ions—also known as plasma.

At intensities of $\geq 10^{13}$ W/cm^2 only a fraction of the material is being ionized and electrons can easily recombine with the ions. This phenomenon leads to non-linear effects providing many important applications e.g. High-Harmonic generation (HHG) in gases and the generation of attosecond pulses. At higher intensities the interaction of laser pulses with solids creates hot-dense plasmas which can be used to construct X-ray lasers. If the laser intensity is increased further, the border of the relativistic regime will be reached at intensities above 10^{18} W/cm^2. This regime is characterized by relativistic velocities of electrons accelerated in the laser field. In this case relativistic effects and the magnetic component of the laser field cannot be neglected anymore. Electrons as well as protons can be accelerated up to energies of 1 GeV and 58 MeV [1, 2], respectively with laser systems which are available today ($\sim 10^{21}$ W/cm^2). In case of ions and protons, electric fields at the rear side of irradiated solid targets are responsible for the acceleration. They reach field strengths of about 10^{12} V/m with a lifetime of a several picoseconds. These field dynamics are essential for the acceleration process and need to be studied to understand the acceleration mechanisms.

The most pronounced differences to proton beams produced by conventional accelerators are the low emittance (high laminarity) and the short duration of the proton bunches (of the order of a picosecond at the source). Different applications

T. Sokollik, *Investigations of Field Dynamics in Laser Plasmas with Proton Imaging,* Springer Theses, DOI: 10.1007/978-3-642-15040-1_1,
© Springer-Verlag Berlin Heidelberg 2011

established recently benefit from these beam attributes. High-energy-density matter can be created, which is of interest for astrophysics [3, 4]. Furthermore, these beams are predestined for temporally and spatially resolved pump-probe experiments like e.g. Proton Imaging.

Laser induced particle beams have also a high potential for future applications. They could be injected into common accelerators, benefitting from the unique attributes of the beams [4, 5]. Further on, the advantages of laser induced proton beams are discussed in the scope of cancer therapy [6, 7]. Since a proton beam of a certain energy deposits its energy mainly in the Bragg peak, it can be used to destroy tumors in regions which are difficult to access surgically (e.g. eye, cerebric). The appealing idea is to produce and shape ion beams by intense laser light much closer to the final interaction point, since light can be easily reflected by mirrors whereas the ion beam transport in conventional accelerators requires huge magnetic systems. Another possible medical application is the creation of radio-isotopes used in positron emission tomography (PET) [8].

In fact, proton and ion beam parameters which are accessible today are far away from being used in the above mentioned applications. Therefore further investigations of the acceleration mechanisms are required to achieve higher proton energies and tailored proton spectra. The progress in this area of research is growing rapidly and in the future this acceleration scheme could lead to very compact and cost-efficient accelerators which might be also accessible to a broader community than conventional accelerators today.

One possibility to reach these goals is to vary the laser parameters. The most promising parameters are the intensity and the contrast of the laser pulse. Thus, ever more powerful lasers are being built and new techniques for pulse cleaning are being developed [9–11]. If these new laser parameters will be available in the near future they will open a door to further physical processes and to new acceleration schemes.

Another important issue is the choice of the target—the ion source. Recently different target types were investigated to shape the ion beams. By using curved targets the emission angle of the ion beam can be influenced. Concave targets can focus or collimate the whole ion beam [3, 12–14]. To achieve tailored spectra, especially monoenergetic ion beams, different approaches exist. For instance, in reference [15] micro-structured targets were used. In reference [16] quasi-monoenergetic ions are accelerated by heating the target and thus manipulating the target surface. At the Max-Born-Institute it was shown for the first time that water-droplet targets can deliver nearly monoenergetic deuteron and proton beams [17, 18].

In the presented thesis different acceleration scenarios are investigated by using different target types, in order to get a further insight into complex relations between laser-plasma interaction and the associated strong fields. This knowledge about the acceleration processes could help to improve and develop novel acceleration schemes. Beside the interest in these basic physically issues the motivation for this research is to achieve knowledge to answer questions like: How can we built laser-ion accelerators to achieve certain ion energies with tailored spectra most efficiently. Answers can be given with experiments and novel measurement tools which are topics of this work.

A powerful diagnostic tool for these investigations is the proton beam itself. It can be used to investigate the acceleration process by probing fields inside a second laser-induced plasma where proton and ion acceleration takes place. This technique is called "Proton Imaging" or "Proton Radiography" and is used for several investigations presented in this thesis. Laser interactions with thin foils and mass-limited targets (water-droplets) at laser intensities between 10^{17} and 10^{18} W/cm^2 will be discussed. Therefore common proton imaging schemes were adapted and developed further. These novel techniques allow detailed investigations of huge transient electric fields (10^8–10^{12} V/m) responsible for the proton (ion) acceleration and connected to the expansion into the vacuum. Additionally, investigations of the beam characteristics deliver information about the acceleration scenario and are included in this thesis.

References

1. W.P. Leemans, B. Nagler, A.J. Gonsalves, C. Toth, K. Nakamura, C.G.R. Geddes, E. Esarey, C.B. Schroeder, S.M. Hooker, GeV electron beams from a centimetre-scale accelerator. Nat. Phys. **2**, 696 (2006)
2. R.A. Snavely, M.H. Key, S.P. Hatchett, T.E. Cowan, M. Roth, T.W. Phillips, M.A. Stoyer, E.A. Henry, T.C. Sangster, M.S. Singh, S.C. Wilks, A. MacKinnon, A. Offenberger, D.M. Pennington, K. Yasuike, A.B. Langdon, B.F. Lasinski, J. Johnson, M.D. Perry, E.M. Campbell, Intense high-energy proton beams from Petawatt-Laser irradiation of solids. Phys. Rev. Lett. **85**(14), 2945 (2000)
3. P.K. Patel, A.J. Mackinnon, M.H. Key, T.E. Cowan, M.E. Foord, M. Allen, D.F. Price, H. Ruhl, P.T. Springer, R. Stephens, Isochoric heating of solid-density matter with an ultrafast proton beam. Phys. Rev. Lett. **91**(12), 125004 (2003)
4. J. Fuchs, P. Antici, E. d'Humieres, E. Lefebvre, M. Borghesi, E. Brambrink, C.A. Cecchetti, M. Kaluza, V. Malka, M. Manclossi, S. Meyroneinc, P. Mora, J. Schreiber, T. Toncian, H. Pepin, R. Audebert, Laser-driven proton scaling laws and new paths towards energy increase. Nat. Phys. **2**(1), 48–54 (2006)
5. T.E. Cowan, J. Fuchs, H. Ruhl, A. Kemp, P. Audebert, M. Roth, R. Stephens, I. Barton, A. Blazevic, E. Brambrink, J. Cobble, J. Fernandez, J.C. Gauthier, M. Geissel, M. Hegelich, J. Kaae, S. Karsch, G.P. Le Sage, S. Letzring, M. Manclossi, S. Meyroneinc, A. Newkirk, H. Pepin, N. Renard-LeGalloudec, Ultralow emittance, multi-MeV proton beams from a laser virtual-cathode plasma accelerator. Phys. Rev. Lett. **92**(20), 204801 (2004)
6. S.V. Bulanov, T.Z. Esirkepov, V.S. Khoroshkov, A.V. Kunetsov, F. Pegoraro, Oncological hadrontherapy with laser ion accelerators. Phys. Lett. A **299**(2–3), 240–247 (2002)
7. J. Fan, W. Luo, E. Fourkal, T. Lin, J. Li, I. Veltchev, C.M. Ma, Shielding design for a laser-accelerated proton therapy system. Phys. Med. Biol. **52**(13), 3913–3930 (2007)
8. K.W.D. Ledingham, P. McKenna, R.P. Singhal, Applications for nuclear phenomena generated by ultra-intense lasers. Science **300**(5622), 1107–1111 (2003)
9. A. Jullien, O. Albert, F. Burgy, G. Hamoniaux, L.P. Rousseau, J.P. Chambaret, F. Auge-Rochereau, G. Cheriaux, J. Etchepare, N. Minkovski, S.M. Saltiel, 10^{-10} temporal contrast for femtosecond ultraintense lasers by cross-polarized wave generation. Opt. Lett. **30**(8), 920–922 (2005)

10. C. Thaury, F. Quere, J.P. Geindre, A. Levy, T. Ceccotti, P. Monot, M. Bougeard, F. Reau, P. d/'Oliveira, P. Audebert, R. Marjoribanks, P. Martin, Plasma mirrors for ultrahigh-intensity optics. Nat. Phys. **3**(6), 424–429 (2007)
11. M.P. Kalashnikov, E. Risse, H. Schonnagel, A. Husakou, J. Herrmann, W. Sandner, Characterization of a nonlinear filter for the frontend of a high contrast double-CPA Ti: sapphire laser. Opt. Exp. **12**(21), 5088–5097 (2004)
12. J. Badziak, Laser-driven generation of fast particles. Opto-Electron. Rev. **15**(1), 1–12 (2007)
13. T. Okada, A.A. Andreev, Y. Mikado, K. Okubo, Energetic proton acceleration and bunch generation by ultraintense laser pulses on the surface of thin plasma targets. Phys. Rev. E **74**(2), 5 (2006)
14. S.C. Wilks, A.B. Langdon, T.E. Cowan, M. Roth, M. Singh, S. Hatchett, M.H. Key, D. Pennington, A. MacKinnon, R.A. Snavely, Energetic proton generation in ultra-intense laser-solid interactions. Phys. Plasmas **8**(2), 542–549 (2001)
15. H. Schwoerer, S. Pfotenhauer, O. Jackel, K.U. Amthor, B. Liesfeld, W. Ziegler, R. Sauerbrey, K.W.D. Ledingham, T. Esirkepov, Laserplasma acceleration of quasi-monoenergetic protons from microstructured targets. Nature **439**(7075), 445–448 (2006)
16. B.M. Hegelich, B.J. Albright, J. Cobble, K. Flippo, S. Letzring, M. Paffett, H. Ruhl, J. Schreiber, R.K. Schulze, J.C. Fernandez, Laser acceleration of quasi-monoenergetic MeV ion beams. Nature **439**(7075), 441–444 (2006)
17. S. Ter-Avetisyan, M. Schnurer, P.V. Nickles, M. Kalashnikov, E. Risse, T. Sokollik, W. Sandner, A. Andreev, V. Tikhonchuk, Quasimonoenergetic deuteron bursts produced by ultraintense laser pulses. Phys. Rev. Lett. **96**(14), 145006 (2006)
18. A.V. Brantov, V.T. Tikhonchuk, O. Klimo, D.V. Romanov, S. Ter-Avetisyan, M. Schnurer, T. Sokollik, P.V. Nickles, Quasi-monoenergetic ion acceleration from a homogeneous composite target by an intense laser pulse. Phys. Plasmas **13**(12), 10 (2006)

Part I
Basics

Chapter 2
Ultra Short and Intense Laser Pulses

In the following chapter fundamental aspects of laser pulses and their interaction with single electrons will be discussed. At first the mathematical description of laser pulses is given. The relation between time and frequency domain will be explained and the concept of generating ultra short pulses will be sketched shortly. Then the interaction of the laser pulse with single electrons in the relativistic case will be discussed and the ponderomotive force will be introduced.

2.1 Mathematical Description

The electric field of short laser pulses can be described either in the time or the frequency domain. Both formalisms are related to each other by the Fourier transformation:

$$E(t) = \frac{1}{\sqrt{2\pi}} \int_{-\infty}^{\infty} \tilde{E}(\omega)\, e^{i\omega t}\, d\omega, \tag{2.1}$$

$$\tilde{E}(\omega) = \frac{1}{\sqrt{2\pi}} \int_{-\infty}^{\infty} E(t)\, e^{-i\omega t}\, dt. \tag{2.2}$$

Due to the fact that $E(t)$ is real, the symmetry of $\tilde{E}(\omega)$ is given as follows:

$$\tilde{E}(\omega) = \tilde{E}^*(-\omega), \tag{2.3}$$

where (*) indicates the complex conjugated function. The symmetry shows that the whole information of the pulse is already given in the positive part of the function. Thus, the reduced function $\tilde{E}^+(\omega)$ is defined as:

T. Sokollik, *Investigations of Field Dynamics in Laser Plasmas with Proton Imaging*, Springer Theses, DOI: 10.1007/978-3-642-15040-1_2,
© Springer-Verlag Berlin Heidelberg 2011

$$\tilde{E}^+(\omega) = \begin{cases} \tilde{E}(\omega) & \text{if} \quad \omega \geq 0 \\ 0 & \text{if} \quad \omega < 0 \end{cases} \qquad (2.4)$$

The inverse Fourier transformation of $\tilde{E}^+(\omega)$ delivers a description of the electric field which is a complex function now. Thus, both functions can be expressed by their amplitude and phase:

$$E^+(t) = A_{ampl}(t)\, e^{i\phi(t)}, \qquad (2.5)$$

$$\tilde{E}^+(\omega) = \tilde{A}_{ampl}(\omega)\, e^{-i\tilde{\phi}(\omega)}. \qquad (2.6)$$

The phase functions $\phi(t)$ and $\tilde{\phi}(\omega)$ can be developed by Taylor series:

$$\phi(t) = \sum_{j=0}^{\infty} \frac{a_j}{j!} t^j, \qquad (2.7)$$

$$\tilde{\phi}(\omega) = \sum_{j=0}^{\infty} \frac{\tilde{a}_j}{j!} \omega^j. \qquad (2.8)$$

The coefficients of the zeroth order $(a_0, \tilde{a}_0)$ represent a constant phase, which shifts the carrier wave within the fixed envelope ("carrier-envelope phase"). The first order coefficients $(a_1, \tilde{a}_1)$ shift the pulse in time and in frequency domain, respectively. With the slowly varying envelope approximation (SVEA) [1] a residual phase can be defined where ω_L represents the central laser (angular) frequency:

$$\varphi(t) = \tilde{\phi}(t) - \omega_L t. \qquad (2.9)$$

The time dependent instantaneous (angular) frequency can be defined by:

$$\omega(t) = \frac{d\phi(t)}{dt} = \omega_L + \frac{d\varphi(t)}{dt}. \qquad (2.10)$$

If the instantaneous frequency is constant in time the pulse is called unchirped and represents a bandwidth-limited pulse, the shortest pulse which can be created with a given spectral width. Pulse duration τ_L and spectral bandwidth $\Delta\omega_p$ are connected by:

$$\tau_L \Delta\omega_L \geq c_B, \qquad (2.11)$$

due to the Fourier transformation (Eqs. 2.1 and 2.2). The constant c_B depends on the spectral shape of the pulse (e.g. gaussian: $c_B = 4\ln 2$).

Higher $(j \geq 1)$ orders of the spectral phase are often not temporally constant. That means that the instantaneous frequency is changing in time. If the frequency increases/decreases the pulse is called up-/down-chirped.[1] A pulse with $d\omega(t)/dt = d^2\varphi(t)/dt^2 = a_2 = const$ is called linearly chirped, the frequency is changing

---

[1] Alternative notation: positive-/negative-chirped.

linearly in time. Pulses gain higher orders of the spectral phase, e.g. by dispersion, when propagating through material. In the Chirped Pulse Amplification (CPA) [2] scheme a linear chirp is generated by different propagation distances of the spectral components. The stretched pulse is amplified without the risk of damaging the optical components (especially the amplifier crystals). After amplification the chirp is compensated by a compressor consisting of two gratings mostly. Alternative schemes exist to compensate higher orders (≥ 1), for example chirped mirrors [3], prisms or a deformable mirror in the compressor to vary the propagation length of the spectral components.

Aside from generating bandwidth-limited pulses by flattening the spectral phase a defined manipulation of the spectral components by a "pulse shaper" (e.g. a liquid-crystal display in the spectral split beam) leads to special temporally shaped pulses which are of interest for several applications [4, 5].

Assuming a constant phase the pulse duration is limited by the spectral bandwidth. To shorten the pulse duration further the bandwidth has to be increased. For this, different methods can be used e.g. self-phase modulation (SPM) in gas-filled hollow fibers [6], or the generation of pulse filaments in gas-filled tubes [7]. The use of these techniques is limited to several mJ pulse energy. Self-phase modulation affects the phase and broadens the bandwidth without influencing the temporal amplitude. Thus, an additional pulse compression is necessary. Recent experiments at the Max-Born-Institute showed that under some conditions the pulse can be self-compressed by pulse filamentation [8, 9]. Details concerning these experiments, including measurements with the MBI TW laser can be found in reference [9].

Using Eqs. 2.5 and 2.9 the electric field can be split into the complex envelope function $A_{ampl}(t)\, e^{i\varphi(t)}$ and the fast oscillating term $e^{i\omega_L t}$:

$$E^+(t) = A_{ampl}(t)\, e^{i\varphi(t)} \cdot e^{i\omega_L t}. \tag{2.12}$$

The real valued temporal electric field can be now reconstructed from $E^+(t)$ as follows:

$$E(t) = 2Re(E^+(t)) \tag{2.13}$$

$$E(t) = 2A_{ampl}(t)\cos(\omega_L t + \varphi(t)). \tag{2.14}$$

Averaging the electric field, the temporal intensity can be calculated:

$$I(t) = \varepsilon_0 c\, \frac{1}{T} \int_{t-T/2}^{t+T/2} E^2(t')\, dt'. \tag{2.15}$$

If the slowly varying envelope approximation is valid Eq. 2.15 can be reduced to:

$$I(t) = 2\varepsilon_0 c A_{ampl}^2(t). \tag{2.16}$$

This formula defines the temporal intensity for a linear polarized laser pulse. In the experiments the peak intensity is usually used to characterize the laser pulses:

$$I_0 = \frac{1}{2}\varepsilon_0 c E_0^2 \tag{2.17}$$

and is typically given in [W/cm^2].

2.2 Single Electron Interaction

The motion of an electron caused by an electromagnetic field E and B in vacuum is described by the Lorentz equation [1]:

$$\frac{dp}{dt} = \frac{d\left(\gamma m_e v\right)}{dt} = -e(E + v \times B). \tag{2.18}$$

In the non-relativistic regime ($\gamma = 1/\sqrt{1 - v^2/c^2} \approx 1$) the electron oscillates in a linearly polarized laser field with an amplitude (y_0) and a maximum velocity (v_0) of:

$$y_0 \approx \frac{e\, E_0}{m_e\, \omega_L^2}, \qquad v_0 \approx \frac{e\, E_0}{m_e\, \omega_L}, \tag{2.19}$$

assuming a plane wave ($E_0 = 2\, A_{ampl}$) with the (angular) frequency ω_L. The maximum velocity is used to define the dimensionless normalized vector potential a_0:

$$a_0 = \frac{v_0}{c} = \frac{e\, E_0}{m_e\, \omega_L\, c}. \tag{2.20}$$

Using Eqs. 2.17 and 2.20 the laser intensity is given by:

$$I_0 = \frac{a_0^2}{\lambda^2} \cdot \frac{\varepsilon_0 m_e^2 c^5}{2e^2}(2\pi)^2 \approx \frac{a_0^2}{\lambda^2} \cdot 1.37 \cdot 10^{18}\ \text{W/cm}^2 \cdot \mu\text{m}^2. \tag{2.21}$$

To calculate the electron trajectories for $a_0 \geq 1$, the equation of motion (Eq. 2.18) has to be discussed fully relativistically. This corresponds to an intensity of $>10^{18}$ W/cm^2. Assuming a linearly polarized plane wave propagating in the x-direction with the vector potential A:

$$A = (0, A_0 \sin\left(k x - \omega_L t\right), 0), \tag{2.22}$$

the electric and magnetic field can be described as follows:

$$E = -\frac{\partial A}{\partial t} = E_0 \cos\left(k x - \omega_L t\right) e_y, \quad E_0 = \omega_L A_0 \tag{2.23}$$

$$\boldsymbol{B} = \nabla \times \boldsymbol{A} = B_0 \cos{(kx - \omega_L t)}\,\boldsymbol{e}_z, \qquad B_0 = kA_0. \tag{2.24}$$

Substituting Eq. 2.18 with these formulas the equation of motion can be written as:

$$\frac{d\boldsymbol{p}}{dt} = e\left(\frac{\partial \boldsymbol{A}}{\partial t} - \boldsymbol{v} \times (\nabla \times \boldsymbol{A})\right). \tag{2.25}$$

The momentum in y-direction is determined by:

$$\frac{dp_y}{dt} = e\left(\frac{\partial A_y}{\partial t} + v_x \frac{\partial A_y}{\partial x}\right) \tag{2.26}$$

$$p_y - eA_y = \alpha_1. \tag{2.27}$$

The constant α_1 is related to the initial momentum in y-direction. Since the magnetic field has a component in z-direction only, the momentum in x-direction is determined by (cf. Eq. 2.18):

$$\frac{dp_x}{dt} = -ev_y B_z = m_e c \frac{d\gamma}{dt}. \tag{2.28}$$

Using the energy equation and the relation between E_0 and B_0 from Eqs. 2.23 and 2.24:

$$\frac{d}{dt}(\gamma m_e c^2) = -e(\boldsymbol{v} \cdot \boldsymbol{E}), \tag{2.29}$$

$$\frac{E_0}{c} = B_0, \tag{2.30}$$

a description of the momentum in x-direction can be found:

$$\gamma - \frac{p_x}{m_e c} = \alpha_2, \tag{2.31}$$

where α_2 is the second invariant of the electron motion. Using $\gamma^2 = 1 + p^2/(m_e c)^2$ a relation between p_x and p_y is given by:

$$\gamma^2 = 1 + \frac{p_x^2}{(m_e c)^2} + \frac{p_y^2}{(m_e c)^2}, \tag{2.32}$$

$$\frac{p_x}{m_e c} = \frac{1 - \alpha_2^2 + (p_y/m_e c)^2}{2\alpha_2}. \tag{2.33}$$

To calculate the electron trajectories the following formula has to be integrated using $\boldsymbol{r} = (x, y, z)$:

$$\boldsymbol{p} = \gamma m_e \frac{d\boldsymbol{r}}{dt} = \gamma m_e \frac{d\boldsymbol{r}}{d\phi}\frac{d\phi}{dt} = -m_e \omega_L \frac{d\boldsymbol{r}}{d\phi}. \tag{2.34}$$

In case of a linearly polarized plane wave in y-direction the trajectories are determined with the initial values $t = 0$, $p_y = 0$, $x = 0$ and $y = 0$ ($\alpha_1 = 0$ and $\alpha_2 = 1$) by:

$$x = \frac{c\, a_0^2}{4\,\omega_L}\left(\phi - \frac{1}{2}\sin(2\phi)\right), \tag{2.35}$$

$$y = \frac{c\, a_0}{\omega_L}(1 - \cos\phi). \tag{2.36}$$

The y-coordinate is oscillating as in the non-relativistic regime, whereas the x-coordinate is oscillating with twice the laser frequency and with an additionally drift. The drift velocity v_D can be estimated averaging over one laser cycle:

$$\bar{x} = \frac{c\, a_0^2}{4\,\omega_L}\phi = \frac{c\, a_0^2}{4\,\omega_L}\,\omega_L\, t - \frac{c\, a_0^2}{4\,\omega_L}\frac{\omega_L}{c}\,\bar{x} \tag{2.37}$$

$$\bar{x} = \frac{c\, a_0^2}{4}\, t - \frac{a_0^2}{4}\,\bar{x} \tag{2.38}$$

$$v_D = \frac{\bar{x}}{t} = \frac{c\, a_0^2}{4 + a_0^2}. \tag{2.39}$$

In Fig. 2.1a, the trajectory of an electron is shown being in rest before hit by a plane wave with infinite duration. In a more realistic case when the electron is deflected by a laser pulse of finite duration, the electron is at rest after the electric field disappears. Thus, the particle does not gain net energy. The electron trajectory is shown in Fig. 2.1b–d for this case. The electron is pushed in laser forward direction while oscillating with the laser frequency and is at rest after the laser pulse is gone. In fact, this case is not realistic at all since the laser pulse is usually focused tightly. If the focus is in the range of the lateral deflection the electron can escape due to the decreasing intensity and thus due to the decreasing restoring force. The electron leaves at an angle θ dependent on its kinetic energy. This phenomenon is called ponderomotive scattering and will be discussed in the next section.

2.3 Ponderomotive Force

Averaging over the equation of motion in time leads to the definition of the ponderomotive force. This force is caused by the gradient of laser intensity which becomes relevant if e.g. a focused laser pulse or a density profile is present. In the following the ponderomotive force in vacuum will be discussed.

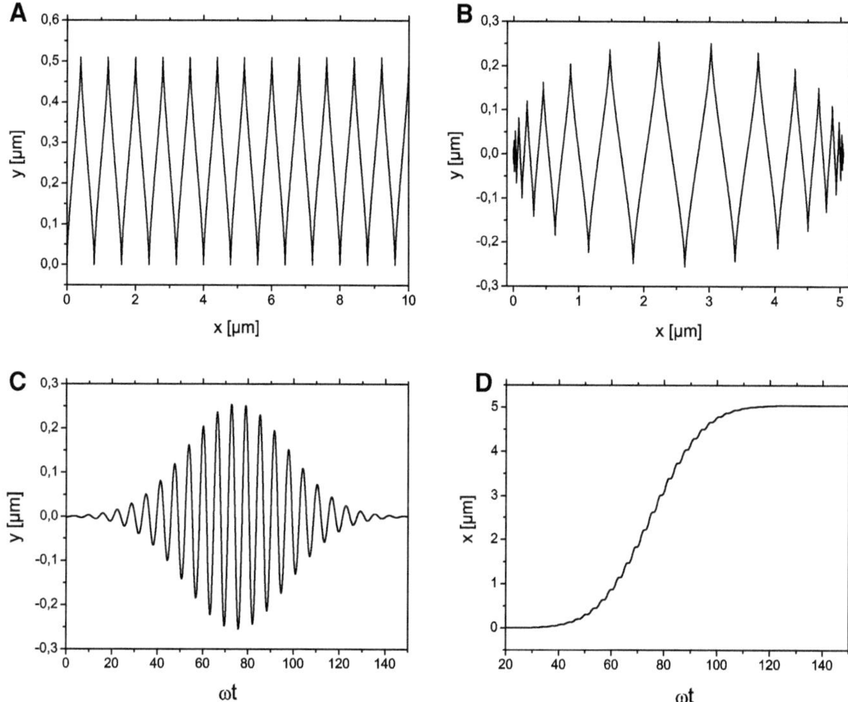

Fig. 2.1 a Electron trajectory caused by a infinite plane wave ($a_0 = 2$) (laboratory frame).
b–d Electron trajectories for a pulse duration of 15 fs with same maximum intensity

In the non-relativistic case ($v/c \ll 1$) the equation of motion can be written as:

$$\frac{\partial v_y}{\partial t} = -\frac{e}{m_e} E_y(r).$$
(2.40)

The electric field E_y, polarized in y-direction and propagating in x-direction, has a radial intensity dependence[2] and can be expressed by a Taylor expansion as follows [1]:

$$E_y(r) \simeq E_0(y) \cos \phi + y \frac{\partial E_0(y)}{\partial y} \cos \phi + \cdots$$
(2.41)

where $\phi = \omega t - k\,x$. The first order can be calculated by integrating Eq. 2.40 using Eq. 2.41:

$$v_y^{(1)} = -\frac{eE_0}{m_e \omega} \sin \phi, \qquad y^{(1)} = \frac{eE_0}{m_e \omega^2} \cos \phi.$$
(2.42)

---

[2] Only the dependence in the y-direction will by considered in the following.

Using Eqs. 2.42 and 2.40 one gets:

$$\frac{\partial v_y^{(2)}}{\partial t} = -\frac{e^2}{m_e^2 \omega^2} E_0 \frac{\partial E_0(y)}{\partial y} \cos^2 \phi. \tag{2.43}$$

Averaging the corresponding force over one cycle leads to:

$$f_{pond} = m_e \overline{\frac{\partial v_y^{(2)}}{\partial t}} = -\frac{e^2}{4m_e \omega^2} \frac{\partial E_0^2(y)}{\partial y}. \tag{2.44}$$

This is the definition of the ponderomotive force in the non-relativistic case. Since the force is dependent on the gradient of E_0^2 electrons will be pushed away from regions of higher intensities. The fully relativistic discussion delivers an additional factor $(1/\langle \gamma \rangle)$ where $\langle \gamma \rangle$ is the relativistic factor γ averaged over the fast oscillations [10]:

$$f_{pond,rel} = -\frac{e^2}{4m_e \langle \gamma \rangle \omega^2} \frac{\partial E_0^2(y)}{\partial y}. \tag{2.45}$$

The angle between electron trajectory and the laser axis can be determined by the ratio of the transversal and longitudinal momentum:

$$\tan \theta = \frac{p_y}{p_x} = \sqrt{\frac{2}{\gamma - 1}}, \tag{2.46}$$

or

$$\cos \theta = \sqrt{\frac{\gamma - 1}{\gamma + 1}}. \tag{2.47}$$

For a linearly polarized and focused laser pulse one would expect an angular spread of the scattered electrons only in the x–y-plane (polarization plane). Due to the fact that for a focused laser pulse an axial magnetic field $B_x = \frac{\partial A_y}{\partial z}$ exists, a force in the z-direction of the same order as the y-component of the ponderomotive force will act. Thus, the electrons are scattered radially symmetrical [1, 11–14].

References

1. P. Gibbon, *Short Pulse Laser Interactions with Matter: An Introduction* (World Scientific Publishing Company, Singapore, 2005)
2. D. Strickland, G. Mourou, Compression of amplified chirped optical pulses. Opt. Commun. **56**(3), 219–221 (1985)
3. R. Szipocs, K. Ferencz, C. Spielmann, F. Krausz, Chirped multilayer coatings for broad-band dispersion control in femtosecond lasers. Opt. Lett. **19**(3), 201–203 (1994)

4. A. Assion, T. Baumert, M. Bergt, T. Brixner, B. Kiefer, V. Seyfried, M. Strehle, G. Gerber, Control of chemical reactions by feedbackoptimized phase-shaped femtosecond laser pulses. Science **282**(5390), 919–922 (1998)
5. T. Brixner, G. Krampert, T. Pfeifer, R. Selle, G. Gerber, M. Wollenhaupt, O. Graefe, C. Horn, D. Liese, T. Baumert, Quantum control by ultrafast polarization shaping. Phys. Rev. Lett. **92**(20), 208301 (2004)
6. M. Nisoli, S. De Silvestri, O. Svelto, Generation of high energy 10 fs pulses by a new pulse compression technique. Appl. Phys. Lett. **68**(20), 2793–2795 (1996)
7. A. Couairon, M. Franco, A. Mysyrowicz, J. Biegert, U. Keller, Pulse self-compression to the single-cycle limit by filamentation in a gas with a pressure gradient. Opt. Lett. **30**(19), 2657–2659 (2005)
8. G. Stibenz, N. Zhavoronkov, G. Steinmeyer, Self-compression of millijoule pulses to 7.8 fs duration in a white-light filament. Opt. Lett. **31**(2), 274–276 (2006)
9. S. Skupin, G. Stibenz, L. Berge, F. Lederer, T. Sokollik, M. Schnurer, N. Zhavoronkov, G. Steinmeyer, Self-compression by femtosecond pulse filamentation: Experiments versus numerical simulations. Phys. Rev. E **74**(5), 056604–9 (2006)
10. B. Quensel, P. Mora, Theory and simulation of the interaction of ultraintense laser pulses with electrons in vacuum. Phys. Rev. E **58**(3), 3719–3732 (1998)
11. G. Malka, J. Fuchs, F. Amiranoff, S.D. Baton, R. Gaillard, J.L. Miquel, H. Pepin, C. Rousseaux, G. Bonnaud, M. Busquet, L. Lours, Suprathermal electron generation and channel formation by an ultrarelativistic laser pulse in an underdense preformed plasma. Phys. Rev. Lett. **79**(11), 2053–2056 (1997)
12. E. Lefebvre, G. Malka, J.L. Miquel, Lefebvre, Malka, and Miquel Reply. Phys. Rev. Lett. **80**(6), 1352 (1998)
13. K.T. McDonald, Comment on "Experimental observation of electrons accelerated in vacuum to relativistic energies by a high-intensity laser". Phys. Rev. Lett. **80**(6), 1350 (1998)
14. P. Mora, B. Quesnel, Comment on "Experimental Observation of Electrons Accelerated in Vacuum to Relativistic Energies by a High- Intensity Laser". Phys.l Rev. Lett. **80**(6), 1351 (1998)

Chapter 3
Plasma Physics

In Chap. 2 the interaction of laser light with matter was discussed for single electrons only. Since the plasma consists of a high number of electrons and ions, processes in plasmas are better described by a fluid model. Thus, also collective effects can be treated analytically. In the following chapter a relevant selection of these effects are discussed.

3.1 Light Propagation in Plasmas

If the laser field displaces electrons from ions the charge separation causes a restoring force. Due to their higher mass, ions can be regarded as an immobile charged background. The resonance frequency of the resulting oscillation is called plasma frequency ω_P and is determined by [1]:

$$\omega_p = \sqrt{\frac{e^2 n_e}{\varepsilon_0 m_e}}, \tag{3.1}$$

where n_e is the electron density. For an underdense plasma where $\omega_P < \omega_L$ light can propagate through the plasma. If the plasma frequency increases (increase of the electron density) up to $\omega_P = \omega_L$ the laser light is no longer transmitted (cf. Eq. 3.4). Thus, the critical density n_c is defined by:

$$n_c = \frac{\varepsilon_0 m_e \omega_L^2}{e^2}. \tag{3.2}$$

Using the dispersion relation for electromagnetic waves in a plasma [2]:

$$\omega_L^2 = k^2 c^2 + \omega_P^2, \qquad k^2 = \frac{\omega_L^2 \varepsilon}{c^2}, \tag{3.3}$$

the refractive index ($n_R = \sqrt{\varepsilon}$) of the plasma can be calculated as follows:

T. Sokollik, *Investigations of Field Dynamics in Laser Plasmas with Proton Imaging*, Springer Theses, DOI: 10.1007/978-3-642-15040-1_3, © Springer-Verlag Berlin Heidelberg 2011

$$n_R = \sqrt{1 - \frac{\omega_P^2}{\omega_L^2}} = \sqrt{1 - \frac{n_e}{n_c}}. \tag{3.4}$$

Now it is obvious that for $\omega_P > \omega_L$ the refractive index becomes imaginary and thus the light can not propagate in the overdense (or overcritical) plasma.

This effect is essential for many applications e.g. for optical probing of plasmas where the critical density defines how deep one can look inside the plasma. Therefore usually the second or third harmonic is used to get a slightly deeper look inside [3]. Nevertheless the critical density represents the limiting factor for optical investigations of dense plasmas. In contrast to that, proton imaging presented in Part III is not restricted by this phenomenon and is therefore a powerful tool for plasma investigations.

The reflection of laser light at the critical (density) surface can also be used for applications. Recently the development of plasma mirrors for the temporal pulse cleaning has gained attention [4–8]. Hence, the laser beam is weakly focused on an antireflection coated glass plate so that the peak intensity reaches about 10^{15}–10^{16} W/cm^2. Since the glass plate is transparent, the low intense part in front of the laser peak is transmitted through. If the laser intensity reaches the ionization barrier (10^{12}–10^{13} W/cm^2) a plasma is created and the high intense part of the laser pulse is reflected at the critical surface. With such a device the contrast of the laser pulse can be increased by 2–3 orders of magnitude usually to a value of $1{:}10^{10}$. More details especially of the plasma mirror installed at the MBI can be found in Ref. [9].

Up to now the non-relativistic case $a_0 \ll 1$ was discussed. If the laser intensity increases up to $a_0 \approx 1$ the plasma frequency and the critical density gain an additional term:

$$\omega_{P,rel} = \sqrt{\frac{n_e e^2}{\varepsilon_0 \bar{\gamma} m_e}}, \qquad n_{c,rel} = \frac{\bar{\gamma} \varepsilon_0 m_e \omega_L^2}{e^2}, \tag{3.5}$$

where $\bar{\gamma} = \sqrt{1 + a_0^2/2}$ [10] represents the relativistic mass increase of the electrons which is averaged over the fast oscillation of the laser field and over a large number of electrons. The plasma frequency is now dependent on the laser intensity and thus new phenomena appear.

One effect is the *laser induced transparency*. Equation 3.5 shows that for increasing intensities the plasma frequency decreases. The plasma becomes transparent if:

$$n_e < \frac{\varepsilon_0 \omega_L^2 m_e}{e^2} \cdot \sqrt{1 + \frac{a_0^2}{2}}. \tag{3.6}$$

The laser pulse can penetrate deeper into the plasma than the classical critical density would allow.

Regarding a focused laser beam with the spatial dependence $a(r) = a_0 \exp\left(-r^2/2\sigma_0^2\right)$ the refractive index can be calculated by substituting Eq. 3.4:

$$n_{R,rel} = \sqrt{1 - \frac{\omega_P^2}{\omega_L^2 \sqrt{1 + \frac{a(r)^2}{2}}}}. \tag{3.7}$$

The relativistic refractive index depends on the distance to the beam axis. For a beam profile which peaks at the beam axis ($dn_{R,\ rel}/dr < 0$) it acts like a focusing lens. This phenomenon is called *relativistic self-focusing*. If the critical power of $P_{c,rel} \approx 17.4\,\text{GW} \cdot n_c/n_e$ [11] is reached the laser is focused due to the mass increase of the relativistic electrons. Additionally the mechanism of ponderomotive scattering discussed above pushes the electrons out of the focus (regions of higher intensities). Thus, the electron density is modified dependent on the distance to the beam axis ($n_e(r)$). This leads to an additionally focusing (*ponderomotive self-focusing*) and hence a relativistic plasma channel can be formed [11].

For the same reason, the increase of the electron mass, a similar effect can be observed regarding the temporal beam profile. The intense part of the pulse propagates with a higher group velocity than the less intense part of the beam since the group velocity is defined by:

$$v_g = c \cdot n_{R,rel} = c \cdot \sqrt{1 - \frac{\omega_P^2}{\omega_L^2 \sqrt{1 + \frac{a(t)^2}{2}}}}. \tag{3.8}$$

This results in a temporal *relativistic profile steepening* at the front of the laser pulse (cf. Fig. 3.1).

Apart from relativistic effects discussed above the interaction of a non-relativistic laser beam with a non-linear medium can lead to similar effects.

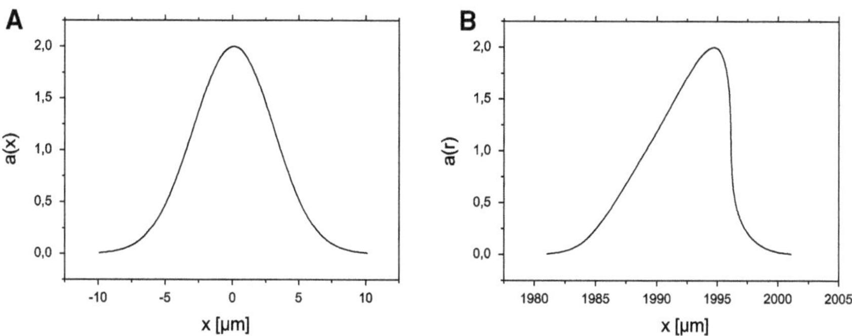

Fig. 3.1 **a** Profile of a 20 fs pulse with $a_0 = 2$. **b** Pulse profile after roughly 2 mm propagation in a medium with $n_e = 1.5 \times 10^{19}\,\text{cm}^{-3}$ whereas only the initial intensity distribution and resulting group velocity was taken into account

The reason therefore is not the increase of the electron mass but the intensity dependent refractive index of non-linear media. Hence phenomena like self-focusing, self-steepening and filamentation can appear [12]. Additionally, self-phase modulation, frequency mixing and other effects were observed and are used in many applications [13–16].

3.2 Debye Length

Inside plasmas a Coulomb potential is shielded by the surrounding electrons. For the discussion the electrons are regarded as a fluid and c.g.s. units are used. Therefore the equation of motion can be written as [2]:

$$n_e e E + \nabla p_e = 0, \tag{3.9}$$

where E is the electric field and p_e the electronic pressure. For an ideal gas the pressure is given by:

$$p_e = n_e k_B T_e, \quad E = -\nabla \varphi_{el}, \tag{3.10}$$

where T_e is the electron temperature and φ_{el} the electrostatic potential. Thus, Eq. 3.9 can be written as:

$$n_e e \nabla \varphi_{el} = k_B T_e \nabla n_e. \tag{3.11}$$

The solution of this equation delivers the electron distribution:

$$n_e = n_{0e} \exp\left(\frac{e\varphi_{el}}{k_B T_e}\right). \tag{3.12}$$

n_{0e} represents the initial electron density. The potential φ_{el} can be determined using the Poisson equation:

$$\nabla^2 \varphi_{el} = -4\pi Z e \delta(r) + 4\pi e (n_e - n_{0e}), \tag{3.13}$$

where $\delta(r)$ is the Dirac function. For $\exp\left(\frac{e\varphi_{el}}{k_B T_e}\right) \ll 1$ one can assume $\exp\left(\frac{e\varphi_{el}}{k_B T_e}\right) \approx 1 + \frac{e\varphi_{el}}{k_B T_e}$. Substituting Eq. 3.12 into Eq. 3.13 one can write:

$$\left(\nabla^2 - \frac{1}{\lambda_D^2}\right)\varphi_{el} + 4\pi Z e \delta(r) = 0, \tag{3.14}$$

where λ_D is the electron Debye length defined by:

$$\lambda_D = \sqrt{\frac{k_B T_e}{4\pi e^2 n_{0e}}}. \tag{3.15}$$

The solution of Eq. 3.14 is given by:

$$\varphi_{el} = \frac{Ze}{r} \exp\left(-\frac{r}{\lambda_D}\right).$$ (3.16)

This formula shows that the potential is shielded by the electrons depending on their energy and initial density. The effective range of the potential is of the order of the Debye length. The number of electrons in the so-called Debye sphere can be estimated by:

$$N_D = \frac{4}{3}\pi n_e \lambda_D^3,$$ (3.17)

and should be $\gg 1$ for an effective shielding [2].

3.3 Plasma Expansion

The expansion of a plasma, described by an electron population with a Boltzmann-like temperature distribution, determined by T_e and an initial density n_{e0} can be discussed self-similarly for the simplest case starting with a step-like initial ion density (cf. Fig. 3.2). It is assumed that the electron distribution is in thermal equilibrium with the resulting electrostatic potential φ_{el}:

$$n_e = n_{e0} \exp\left(\frac{e\varphi_{el}}{k_B T_e}\right).$$ (3.18)

Using the Poisson equation and $Z\,n_{i0} = n_{e0}$ for $x \leq 0$ as well as $\varphi_{el}(-\infty) = 0$ one obtains:

$$\varepsilon_0 \frac{\partial^2 \varphi_{el}}{\partial x^2} = e(n_e - Zn_i).$$ (3.19)

This can be integrated analytically for $x = 0$ to $x = \infty$ and delivers the initial electric field at $x = 0$ [19]:

$$E_{Front,0} = \sqrt{\frac{2}{e_E}} E_i,$$ (3.20)

where $E_i = \sqrt{n_{e0} k_B T_e / \varepsilon_0}$ and $e_E = 2.71828\ldots$ This initial electric field depends on the initial electron density n_{e0} and the electron temperature T_e only. It results from the charge separation at the surface and thus from the leaking of the hot electrons. The field is responsible for ionization of the contamination layer at the rear surface of laser irradiated targets and for ion acceleration. For an initial electron density of $n_{e0} = 5 \times 10^{20}$ cm^{-3} and a temperature of $T_e = 600$ keV the electric field is about 2×10^{12} V/m (see also Chap. 4).

The temporal evolution of the ion and electron densities is described by the equations of continuity and motion [19]:

$$\left(\frac{\partial}{\partial t} + v_i \frac{\partial}{\partial x}\right) n_i = -n_i \frac{\partial v_i}{\partial x}, \tag{3.21}$$

$$\left(\frac{\partial}{\partial t} + v_i \frac{\partial}{\partial x}\right) v_i = -\frac{Ze}{m_i} \frac{\partial \varphi_{el}}{\partial x}, \tag{3.22}$$

where v_i is the ion velocity and m_i the ion mass. For $x + c_s\, t > 0$, with the ion-acoustic velocity $c_s = \sqrt{Zk_B T_e/m_i}$, a self-similar solution exist, if quasi-neutrality in the expanding plasma, $n_e = Zn_i = n_{e0} \exp\left(-x/c_s t - 1\right)$ is assumed. The ion velocity v_i and the self-similar electric field E_s is then given by:

$$v_i = c_s + \frac{x}{t}, \tag{3.23}$$

$$E_s = \frac{k_B T_e}{e c_s t} = \frac{E_i}{\omega_{pi} t}, \tag{3.24}$$

where $\omega_{pi} = \sqrt{n_e Z e^2/\varepsilon_0 m_i}$ is the ion plasma frequency. The self-similar solution is not valid for $\omega_{pi} t < 1$ when the initial Debye length is larger than the self-similar scale length $c_s t$. Also for $\omega_{pi} t \gg 1$ the ion velocity would increase unlimited for $x \to \infty$. The self-similar solution becomes invalid when the Debye length equals the density scale length $c_s t$, this position corresponds to the ion front [19]:

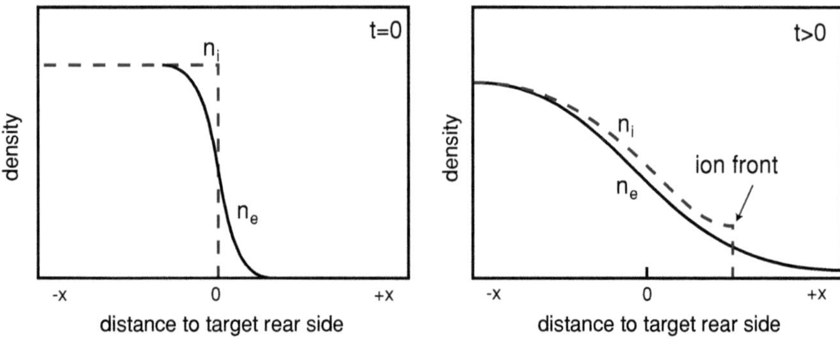

Fig. 3.2 On the *left hand side* the initial density distributions are shown schematically ($t = 0$). The electrons leak into the vacuum creating an electric field which drives the plasma expansion. On the *right hand side* the densities are plotted during the expansion process ($t > 0$). Note: The shown densities are sketching the results of numerical simulations [17, 18] and can not be achieved by the self similar solution

$$\lambda_D = \lambda_{D0} \cdot \sqrt{\frac{n_{e0}}{n_e}} \tag{3.25}$$

$$\lambda_D = \lambda_{D0} \cdot \exp\left(\frac{1 + \frac{x}{c_s t}}{2}\right) \tag{3.26}$$

$$\lambda_D \equiv c_s t \tag{3.27}$$

At this position Eq. 3.23 predicts an ion velocity of:

$$v_p = 2c_s \ln \omega_{pi} t. \tag{3.28}$$

Integrating Eq. 3.24 over time, the electric field is determined by [19]:

$$E_{Front,\omega_{pi}t \gg 1}(t) = 2 \frac{E_i}{\omega_{pi} t}. \tag{3.29}$$

Equations 3.20 and 3.29 deliver two solutions for the electric field at the ion front, for $t = 0$ and for $\omega_{pi} t \gg 1$. To describe the electric fields for all times and the temporal evolution of the densities also for $x < 0$, numerical methods have to be applied. In [18, 19] an expression for the field at the front has been found:

$$E_{Front}(t) = \sqrt{\frac{2}{e_E}} \frac{E_i}{\sqrt{1 + \tau_F^2}}, \tag{3.30}$$

with $\tau_F = \omega_{pi} t / \sqrt{2e_E}$. For $t = 0$ and $\omega_{pi} \gg 1$ this expression is similar to the formulas discussed above (see also [17]).

Apart from the electric field which peaks at the ion front E_{Front}, the electric field between target surface and the front is almost homogeneous and decreases with $E_{Plateau} \propto t^{-2}$ [18, 19].

In a simplifying picture the electric field $E_{Front,0}$ accelerates the ions which are expanding together with the electrons into the vacuum with the decreasing electric fields $E_{Front}(t)$ and $E_{Plateau}(t)$. The proton imaging discussed in detail in Part III can visualize the influence of these fields on the proton probe beam. Thus, information about the acceleration and the expanding plasma can be gained by this method.

References

1. P. Gibbon, *Short Pulse Laser Interactions with Matter: An Introduction* (World Scientific Publishing Company, Singapore, 2005)
2. S. Eliezer, *The Interaction of High Power Lasers with Plasmas* (IOP Publishing, Bristol, 2002)
3. E.T. Gumbrell, R.A. Smith, T. Ditmire, A. Djaoui, S.J. Rose, M.H.R. Hutchinson, Picosecond optical probing of ultrafast energy transport in short pulse laser solid target interaction experiments. Phys. Plasmas **5**(10), 3714–3721 (1998)

4. C. Thaury, F. Quere, J.P. Geindre, A. Levy, T. Ceccotti, P. Monot, M. Bougeard, F. Reau, P. d'Oliveira, P. Audebert, R. Marjoribanks, P. Martin, Plasma mirrors for ultrahigh-intensity optics. Nat. Phys. **3**(6), 424–429 (2007)
5. P. Antici, J. Fuchs, E. d'Humieres, E. Lefebvre, M. Borghesi, E. Brambrink, C.A. Cecchetti, S. Gaillard, L. Romagnani, Y. Sentoku, T. Toncian, O. Willi, P. Audebert, H. Pepin, Energetic protons generated by ultrahigh contrast laser pulses interacting with ultrathin targets. Phys. Plasmas **14**(3), 030701 (2007)
6. A. Levy, T. Ceccotti, P. D'Oliveira, F. Reau, M. Perdrix, F. Qurere, P. Monot, M. Bougeard, H. Lagadec, P. Martin, J.P. Geindre, P. Audebert, Double plasma mirror for ultrahigh temporal contrast ultraintense laser pulses. Opt. Lett. **32**(3), 310–312 (2007)
7. D. Neely, P. Foster, A. Robinson, F. Lindau, O. Lundh, A. Persson, C.G. Wahlstrom, P. McKenna, Enhanced proton beams from ultrathin targets driven by high contrast laser pulses. Appl. Phys. Lett. **89**(2), 021502 (2006)
8. G. Doumy, F. Quere, O. Gobert, M. Perdrix, P. Martin, P. Audebert, J.C. Gauthier, J.P. Geindre, T. Wittmann, Complete Characterization of a Plasma Mirror for the Production of High-Contrast Ultraintense Laser Pulses. Phys. Rev. E, **69**(2), 026402 (2004)
9. S. Steinke, *Entwicklung eines Doppel-Plasmaspiegels zur Erzeugung hochenergetischer Ionen mit ultra-dünnen Targets*. Diploma Theses (Freie Universität Berlin, 2007)
10. E. Lefebvre, G. Bonnaud, Transparency/Opacity of a Solid Target Illuminated by an Ultrahigh-Intensity Laser Pulse. Phys. Rev. Lett. **74**(11), 2002–2005 (1995)
11. G.Z. Sun, E. Ott, Y.C. Lee, P. Guzdar, Self-focusing of short intense pulses in plasmas. Phys. Fluids **30**(2), 526–532 (1987)
12. J.K. Ranka, A.L. Gaeta, Breakdown of the slowly varying envelope approximation in the self-focusing of ultrashort pulses. Opt. Lett. **23**(7), 534–536 (1998)
13. R. Butkus, R. Danielius, A. Dubietis, A. Piskarskas, A. Stabinis, Progress in chirped pulse optical parametric amplifiers. Appl. Phys. B **79**(6), 693–700 (2004)
14. T. Kobayashi, A. Shirakawa, Tunable visible and near-infrared pulse generator in a 5 fs regime. Appl. Phys. B **70**, S239–S246 (2000)
15. M. Nisoli, S. DeSilvestri, O. Svelto, R. Szipocs, K. Ferencz, C. Spielmann, S. Sartania, F. Krausz, Compression of high-energy laser pulses below 5 fs. Opt. Lett. **22**(8), 522–524 (1997)
16. T. Brabec, F. Krausz, Intense few-cycle laser fields: frontiers of nonlinear optics. Rev. Mod. Phys. **72**(2), 545–591 (2000)
17. M.C. Kaluza, *Characterisation of Laser-Accelerated Proton Beams*. Ph.D. thesis, Technischen Universität München, 2004
18. P. Mora, Thin-foil expansion into a vacuum. Phys. Rev. E **72**(5), 056401 (2005)
19. P. Mora, Plasma expansion into a vacuum. Phys. Rev. Lett. **90**(18), 185002 (2003)

Chapter 4
Ion Acceleration

Due to their high mass, ions cannot be accelerated directly by the field of currently achievable laser pulses. Substituting the electron mass by the 1,836 times higher proton mass (m_p) in the vector potential a_0, the averaged kinetic energy which can be gained is defined by the ponderomotive potential $\Phi_{\text{pond,prot}}$ as follows [1, 2]:

$$\Phi_{\text{pond,prot}} = m_p c^2 (\gamma - 1) = m_p c^2 \left(\sqrt{1 + \frac{a_0^2}{1,836^2}} - 1 \right). \qquad (4.1)$$

For $a_0 = 3$ $(I \approx 1.5 \times 10^{19}\,\text{W/cm}^2)$ the ponderomotive potential becomes $\Phi_{\text{pond,prot}} \approx 1.3\,\text{keV}$ which can be neglected within the scope of ion acceleration. The relativistic threshold for protons is fulfilled for $a_0 = 1{,}836$ which corresponds to an intensity of about $5 \times 10^{24}\,\text{W/cm}^2$ and cannot be realized by the present laser technology. But, a high fraction of the laser energy can be transferred to electrons by multiple processes. The resulting hot electrons create a charge separation. Thus, huge electrostatic fields are generated accelerating ions to energies of several MeV. In the following several absorption mechanisms are discussed to give an overview of the multifarious and complex physical processes. They all deliver a fraction to the energy absorption whereas the dominance of the mechanisms are determined by the experimental conditions, e.g. temporal contrast and energy of the laser pulse, angle of incidence and target type.

4.1 Absorption Mechanisms

4.1.1 Resonance Absorption

Resonance absorption can take place if a p-polarized laser pulse irradiates the target at an angle θ. The electric field perpendicular to the target surface tunnels

T. Sokollik, *Investigations of Field Dynamics in Laser Plasmas with Proton Imaging,* Springer Theses, DOI: 10.1007/978-3-642-15040-1_4, © Springer-Verlag Berlin Heidelberg 2011

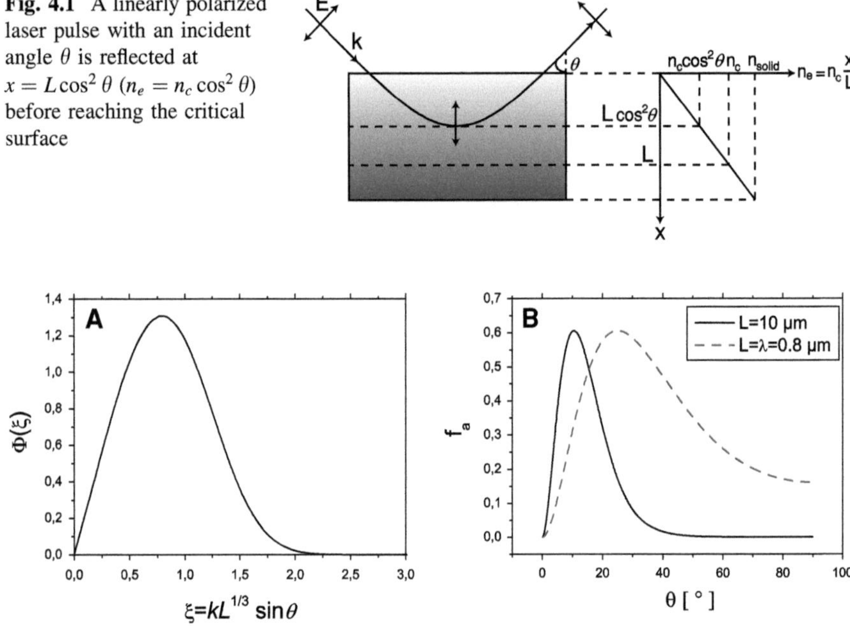

Fig. 4.1 A linearly polarized laser pulse with an incident angle θ is reflected at $x = L\cos^2\theta$ ($n_e = n_c\cos^2\theta$) before reaching the critical surface

Fig. 4.2 **a** The angular dependence of the resonance absorption determined by Eq. 4.2. **b** The absorption for two different plasma scale lengths ($L = 10\,\mu\mathrm{m}$ and $L = \lambda = 0.8\,\mu\mathrm{m}$)

through the critical surface and resonantly drives an electron plasma wave. This wave can be damped either by collision or collisionless processes. Hot electrons are created since only a minority of the plasma electrons acquires most of the absorbed energy. In Fig. 4.1 the density profile with the scale length L is shown where $n_e = n_c x/L$. The scale length depends on the temporal contrast of the laser pulse. The Amplified Spontaneous Emission (ASE, see Sect. 5.1) or pre-pulses create a plasma at the surface of the target. A higher contrast causes a smaller scale length. If the laser pulse irradiates the target at an angle θ the light is reflected before reaching the critical surface at $x = L\cos^2\theta$ (see Fig. 4.1). A small scale length allows the light to tunnel through the critical surface where the plasma wave can be driven resonantly. A formulation of the absorption fraction can be found for large scale length ($kL \gg 1$). The behavior of the angular absorption is given by $\Phi(\xi)$ where $\xi = (kL)^{1/3}\sin\theta$ [3, 4]:

$$\Phi(\xi) \simeq 2.3\xi\exp\left(\frac{-2\xi^3}{3}\right). \tag{4.2}$$

This function is plotted in Fig. 4.2 A and has a maximum for $\xi \approx 0.8$ which corresponds to an angle of $\sin\theta = 0.43(\lambda/L)^{1/3}$. That means that for larger scale lengths this angle is decreasing. For $L = \lambda$ the maximum absorption can be reached at $\theta = 25°$. For a linear density profile the laser absorption f_a can be estimated introducing a moderate damping [1, 5] by:

$$f_a = \frac{I_{abs}}{I_L} \simeq 36\xi^2 \frac{\mathrm{Ai}(\xi)^3}{\left|\frac{d\mathrm{Ai}(\xi)}{d\xi}\right|}, \qquad (4.3)$$

where Ai is the Airy function. At the maximum ($\xi \approx 0.8$) the absorption is about 60%. f_a is plotted in Fig. 4.2 B for two different scale lengths.

4.1.2 Brunel Absorption (Vacuum Heating)

This mechanism is named after F. Brunel who published the famous article "Not-so-resonant, Resonant Absorption" [6] in the year 1987. If the oscillating electrons along the density gradient with the amplitude $x_p \simeq eE_L/m_e\omega_L^2 = v_{os}/\omega_L$ exceed the density scale length $v_{os}/\omega_L > L$ the resonance breaks down [1]. On the other hand electrons close to the edge of a sharp density profile can be accelerated into the vacuum during one laser half cycle. When the field reverses these electrons are accelerated back into the plasma as the laser field cannot penetrate into the overdense plasma. Thus, electrons are accelerated into the target and gain kinetic energy. In this simple model the $v \times B$ term is neglected. The angular dependence of the Brunel absorption can be estimated analytically assuming a step-like density profile and a relativistic increase of the electron mass. More details can be found in references [1, 6]. As a result of the analytical discussion two limiting cases will be presented, for $a_0 \ll 1$ and $a_0 \gg 1$. At low intensities the angular dependence is determined [1] by:

$$\eta_{a_0\ll 1} = \frac{a_0}{2\pi}f^3\alpha(\theta), \qquad (4.4)$$

where θ is the incident angle, $\alpha(\theta) = \sin^3\theta/\cos\theta$, $f = (\sqrt{1+8\beta} - 1)/(2\beta)$ and $\beta = a_0\,\alpha/2\pi$. The function is plotted in Fig. 4.3.

In the strongly relativistic case $a_0 \gg 1$ the angular dependence is defined by:

$$\eta_{a_0\gg 1} = \frac{4\pi\alpha'(\theta)}{(\pi + \alpha'(\theta))^2}, \qquad (4.5)$$

where $\alpha'(\theta) = \sin^2\theta/\cos\theta$. The term is independent on the laser intensity and exhibit a maximum at $\alpha'(\theta) = \pi$ which corresponds to an angle of $\theta = 73°$. Together with the low intensity case this function is plotted in Fig. 4.3.

It has to be taken into account that in this simple model several mechanisms are neglected. For high laser intensities the $v \times B$ term influences the electron trajectories and pushes the electrons in laser forward direction (see Chap. 3). Also a finite density profile should be considered for a more realistic case. In reference [7]

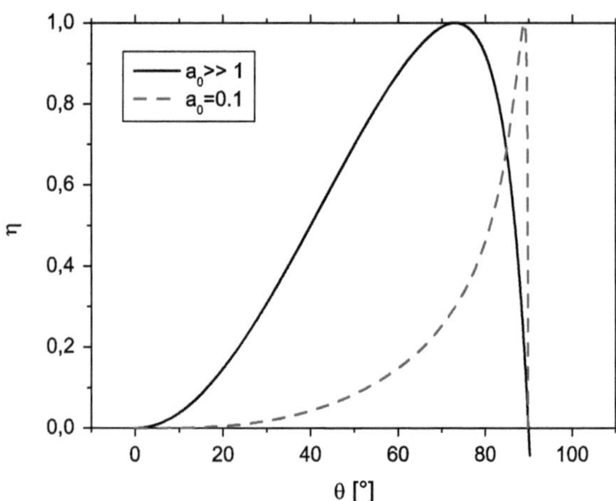

Fig. 4.3 The angular dependence of the Brunel absorption is plotted for $a_0 \gg 1$ and $a_0 = 0.1$

simulations for different plasma scale length were done showing an absorption of the laser energy about 70% at low intensities ($I\lambda^2 = 10^{16}\,\mathrm{W/cm^2\,\mu m^2}$) and scale lengths of $L/\lambda \sim 0.1$. At high intensities and short scale lengths the absorption is only about 10–15%. It has also been shown that the optimum angle is about 45° for such intensities [1] which differs from the estimations discussion above. The reasons for this are DC currents at the target surface which create a magnetic field and inhibit the returning of the electrons to the plasma.

4.1.3 Ponderomotive Acceleration, Hole Boring and j × B Heating

The ponderomotive force, discussed in Sect. 2.3, pushes electrons away from the area of high intensities along the field gradient. The averaged kinetic energy they can gain is given by the ponderomotive potential:

$$\Phi_{\mathrm{pond}} = E_{\mathrm{kin}} = m_e c^2 (\gamma - 1) \tag{4.6}$$

$$\Phi_{\mathrm{pond}} = E_{\mathrm{kin}} = m_e c^2 \left(\sqrt{1 + a_0^2} - 1 \right). \tag{4.7}$$

At an intensity of $a_0 = 3$ the kinetic energy which is usually associated with the electron temperature can be estimated to be about 690 keV. The field gradient is directed normal to the critical surface and thus electrons are accelerated perpendicular to the surface of the overcritical plasma into the target.

In fact, this surface will be modulated by the interaction with the laser pulse. The light pressure $p_L = 2I c$ [1, 2] (for normal incidence and without laser absorption) reaches a value of about 6.6 Gbar at an intensity of 1×10^{19} W/cm^2. Since the laser pulse bores a hole into the plasma this effect is named Hole Boring. The velocity can be estimated by balancing the momentum flux of the mass flow with the light pressure (assuming $p_e \ll p_L$):

$$\frac{\partial}{\partial x}(n_i m_i u u) + \frac{\partial}{\partial x}(Z p_e + p_L) = 0 \qquad (4.8)$$

and using momentum and number conservation ($m_i n_i u = $ const.) [1, 2]:

$$\frac{u}{c} = \sqrt{\frac{n_c Z m_e}{2 n_e m_i} \frac{I \lambda_\mu^2}{1.37 \times 10^{18}}}. \qquad (4.9)$$

u is the velocity of the critical surface, n_i and n_e the ion and electron density and Z the ion charge state. With a ratio of $n_c/n_e = 1/4$ and an intensity of 10^{19} W/cm^2 the velocity has a value of $0.0244\,c$. Thus, the plasma moves roughly 0.4 µm in 50 fs and 14.7 µm in 2 ps. Hence the effect is more pronounced for temporal long pulses with high intensities. This effect was observed experimentally by analyzing the wavelength shift of the reflected laser pulse [8]. The deformation of the plasma influences the directionality of the electrons since they are accelerated along the density gradient. Thus, electrons are not only accelerated perpendicular to the target surface but also in direction between target surface and laser axis.

The $v \times B$ term for relativistic laser intensities causes an oscillation with twice the laser frequency in direction of the laser propagation. This was already discussed in Sect. 2.2 (Eq.2.35). Similar to the Brunel absorption, electrons can gain kinetic energy at a step-like density profile by propagation into the overdense region where no restoring force of the laser light acts. The force on the electrons in laser direction can be written as follows [1]:

$$f_x = -\frac{m}{4}\frac{\partial v_{os}^2(x)}{\partial x}(1 - \cos 2\omega_L t). \qquad (4.10)$$

The first term is the ponderomotive force again (cf. Sect. 2.3) and the second term leads to a heating of the electrons. This mechanism is most effective for normal incidence and high intensities and is named $j \times B$ heating.

4.2 Target Normal Sheath Acceleration

As described above hot electrons are generated due to the coupling of the laser light with the plasma. They gain 10–50% [9–15] of the laser energy and are accelerated in direction of the density gradient or in direction of the laser propagation. They travel through thin targets of (e.g. 13–0.8) µm thickness in a cone of 10°–30° [15] due to deformations of the critical surface by the ponderomotive force. The electrons interact with cold electrons or nuclei inside the target. These collisions cause a scattering of the electron beam and an emission of X-rays and other types of particles [15]. Also collective mechanisms play a role since the current of the electrons is typically in the order of the Alfvén limit, the maximum current that can be transported by an electron beam in vacuum: $I_A(kA) = 17\ \beta\gamma$ [15] or even 2–3 orders of magnitude above. The cold electrons inside the target neutralize the space charge of the beam which otherwise would lead to a coulomb explosion of the beam. On the other hand these cold electrons create a return current which satisfies the Alfvén limit for the total current whereas this limit is exceeded by the hot electron current. The resulting magnetic field around the electron beam guides the propagation of the electrons. Additionally, electro-magnetic and electrostatic instabilities can arise, e.g. the Weibel instability [15–19] and electron filaments can be generated.

When the electrons reach the rear side of the target only the fastest electrons (precursor electrons) can escape since a potential will be build up which hinder the electrons to escape. The electrons which can not escape turn around reentering the target surface and starting to oscillate [20, 21]. These electrons which leak into the vacuum (as discussed in Sect. 3.3) create a so-called electron sheath. This charge separation generate an electric field which accelerates the ions while expanding and decreasing in time.

The initial field at the target rear side reaches values of 10^{12} V/m and above at relativistic laser intensities. This is well above the ionization threshold of the contamination layer which naturally covers the target surfaces. This contamination layer consists of hydrocarbon and water mainly. The positively charged ions are accelerated in this field and are later charge compensated by the electrons which loose energy during the ion acceleration. This scheme called Target Normal Sheath Acceleration (TNSA) is sketched in Fig. 4.4.

The TNSA model was first discussed in reference [14, 22]. The initial ($t = 0$ and $z = 0$) field strength ($E_{x=0}$) was already estimated in Sect. 3.3 by solving the Poisson equation for a step-like ion density and an electron density with T_e being in equilibrium with the potential [14, 23]:

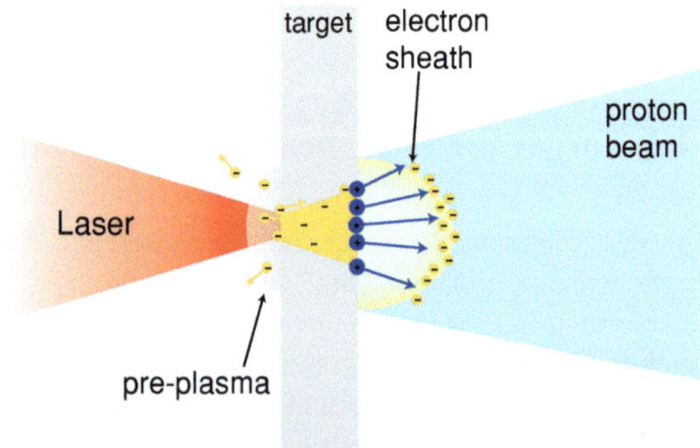

Fig. 4.4 Illustration of the TNSA mechanism. The laser pulse is focused on a target accelerating electrons into the target. These electrons propagate through the target and create an electron sheath at the rear side. Protons are accelerated by the resulting electrostatic field at the rear side of the target

$$E_{x=0} = \sqrt{\frac{2}{e_E}}\sqrt{\frac{n_e k_B T_e}{\varepsilon_0}} \approx \frac{k_B T_e}{e\lambda_D}. \tag{4.11}$$

The kinetic energy of the electrons $(k_B T_e)$ is usually estimated by the ponderomotive potential (Eq. 4.7). The number of hot electrons is determined by the absorbed laser energy E_a which can be estimated from an empirical scaling law [11, 24, 25]:

$$\frac{E_a}{E_L} = \eta \approx a(I_L\lambda^2)^{3/4} \approx 1.68 \times 10^{-15} \times I_L^{3/4}\ (\text{W/cm}^2) \qquad (\lambda = 1\,\mu\text{m}) \tag{4.12}$$

where η has a maximum of 0.5 at an intensity of 3.1×10^{19} W/cm^2. The number of electrons is determined by:

$$N_e \approx \frac{\eta \times E_L}{k_B T_e}. \tag{4.13}$$

The electron density n_e can be estimated by the number of hot electrons occupying a volume defined by the electron spot at the rear side multiplied by the laser pulse duration:

$$n_e \approx \frac{N_e}{c\tau\pi B^2}, \tag{4.14}$$

where $B = r_L + d\tan\theta$, r_L the laser focus, d the target thickness and $\theta \approx 10° - 30°$ the electron propagation angle. Assuming a laser pulse with an intensity of $I_L = 2 \times 10^{19}$ W/cm^2 leads to:

$$\eta \approx 0.35, \qquad T_e \approx 700\ \text{keV}, \qquad \lambda_D \approx 0.14\ \mu\text{m} \tag{4.15}$$

and finally to an initial electric field of about 4×10^{12} V/m.

To estimate the maximal proton energy the isothermal expansion which was discussed in Sect. 3.3 can be used. From the velocity of the ion front for $\omega_{pi} \gg 1$ (cf. Eq. 2.28) the proton energy can be estimated by [23]:

$$E_{\text{prot}} \approx 2k_B T_e \left(\ln\frac{2\omega_{pi}t}{\sqrt{2e_E}}\right)^2, \tag{4.16}$$

where t is in the order of the laser pulse duration τ_L. The best fit to experimental data was found in reference [25] with $t \approx 1.3\tau_L$. For a laser pulse with $I_L = 2 \times 10^{19}$ W/cm^2 and $\tau_L = 40\,\text{fs}$, the maximum proton energy can be estimated:

$$E_{\text{prot}} \approx 1.4\ \text{MeV}. \tag{4.17}$$

In fact, in the experiments higher proton energies (up to 5 MeV) are observed. The acceleration time t seems to be the critical parameter. The scaling with the pulse duration $(t \approx 1.3\tau)$ which was found in reference [25] agrees well with experiments with pulse durations of $\tau_L > 100\,\text{fs}$. But for ultra-short pulses $\tau_L < 100\,\text{fs}$ this scaling can not be used further without any restrictions (cf. [26]).

An alternative possibility to estimate the maximum proton energy was suggested in reference [27]. In contrast to the plasma expansion model (cf. Sect. 3.3) the potential stays finite for $(x \to \infty)$. The ions gain energy by the potential:

$$E_i(z) = -q_i e \Phi(z) = E_{i,\infty}\left(1 + \frac{z}{B} - \sqrt{1 + \left(\frac{z}{B}\right)^2}\right), \tag{4.18}$$

where $E_{i,\infty} = q_i k_B T_e B / \lambda_D$. The maximum proton energy is determined by:

$$\frac{\tau_L}{\tau_0} = X\left(1 + \frac{1}{2(1 - X^2)}\right) + \frac{1}{4}\ln\frac{1 + X}{1 - X}, \tag{4.19}$$

where $\tau_0 = B/(2E_i(\infty)/m_i)^{1/2}$ and $X = (E_{prot}/E_{i,\infty})^{1/2}$. Equation 4.19 predicts a maximum proton energy of $E_{prot} = 2.7\,\text{MeV}$ which fits well with the experimental observations.

It has to be noted that these estimations are very rough and can only deliver the order of magnitude of the parameters. For a better understanding and predictions of the ion beam parameters, PIC simulations or hydrodynamical models are required and can be found in several publications [23, 28–31].

4.3 Alternative Acceleration Mechanisms

The TNSA mechanism discussed above describes only one of the possible accelerations schemes. In recent years a controversy occurs if this (target) "rear-side" process or a "front-side" acceleration is responsible for high energetic (MeV) proton generation. The "front-side" acceleration is explained by the charge separation near the critical surface at the front-side of the target [32]. A similar explanation is given in reference [33] by the so-called shock acceleration. As discussed in Sect. 4.1.3 the laser radiation pressure can accelerate the critical surface forward in laser direction (Hole Boring). This can launch an ion acoustic wave into the target which can evolve into an electrostatic shock [34] which accelerate the ions. One of the arguments for this mechanism is the observed ring structure in the proton beam which can be explained by magnetic fields which arise due to the propagation through the target [33]. But also this feature is discussed controversy since a saturation of the used CR39-detector can explain this structure [35]. On the other hand experiments where the target is pre-heated and thus the contamination layers are removed show a reduced proton signal whereas the ion signal is enhanced [36, 37]. A pre-heated target, coated with CaF_2 at the rear side only was used to identify the "rear-side" acceleration to be responsible for the ion acceleration [36]. Once more it seems that neither the "front-side" nor the "rear-side" can explain all observed phenomena at once. A complex interaction of laser pulse intensity, target thickness and temporal contrast (plasma scale length) causes the dominance of one of the mechanisms. In reference [34] a scenario is discussed where both mechanism can act together to accelerate ions to higher energies. The ions are first accelerated at the front, then they propagate through the target and are accelerated further by the electrostatic field at the rear side. Therefore the intensity has to be high enough to generate fast ions due to the shock. Additionally, the electrostatic field at the rear side has to last long enough so that the ions can still see a strong field when they exit the target rear side [34].

Both mechanisms, shock acceleration and acceleration due to an electrostatic field have been identified in one experiment [38]. The laser pulse was focused on a metal wire at intensities of about $>5 \times 10^{19}$ W/cm^2. A directed proton beam in laser forward direction was observed and attributed to shock-acceleration. Additionally, a disc-like emission was observed and explained by charge effects and thus by the cylindrical electrostatic field at the wire surface. Even an other wire, placed 250 µm away from the illuminated one was charged up and emitted protons isotropically. This experiment is very interesting especially in the scope of the experiments described in Chap. 11 where spherical targets were illuminated showing a similar behavior. Immediately the question arises if the directed proton emission is caused by shock acceleration or due to the special geometry of the target allowing electrons to propagate around the target enhancing the field at the rear side. For the experiments presented in Chap. 11 the intensities were much lower and therefore the shock acceleration can not account for an effective proton acceleration. Hence not only different absorption mechanisms have to be considered also geometrical attributes are relevant for the discussion of the ion acceleration schemes.

Starting from a very simple description of the ion acceleration mechanism (TNSA) in the section above a closer look at the processes shows the complexity of the physics. It also distinguishes the great potential of learning from new effects which appears by reaching new parameter regimes. Thus, the evolution of the scientific knowledge is strongly connected with the development of the laser systems and novel experimental setups (e.g. Plasma Mirrors, manipulated target systems or mass-limited targets). The understanding of the processes is rising rapidly, e.g. in the last 2 years new phenomena like mono-energetic ion and proton bunches (see Sect. 6.2) or laser induced micro-lenses [39] have been observed which have a high potential for further applications. Novel temporal pulse cleaning methods (cf. Sect. 5.1) allow to irradiate targets of only several tens of nanometer today which is connected to new physical effects.

In the so-called transparent regime for example, a part of the laser pulse can be transmitted through thin (<1 µm) targets causing a superior electron heating and proton acceleration at the target rear surface [34]. Effects like relativistic transparency [34] or Coulomb explosion [40] become important. Illuminating even thinner targets (tens of nanometers) with circularly polarized, intense laser pulses with a high contrast, reference [41] predicts the generation of highly energetic and mono-energetic ions. This mechanism is connected to the radiation pressure of the laser light [42, 43].

In summary, the tendency of actual research is directed to this regime—thinner targets or mass-limited targets irradiated by ultra-high intensity laser pulses ($\geq 10^{20}$ W/cm^2) with a contrast of $\geq 10^{10}$. As shortly sketched above this regime demands an advanced description of the acceleration processes and further investigations.

References

1. P. Gibbon, in *Short Pulse Laser Interactions with Matter: An Introduction* (World Scientific Publishing, Singapore, 2005)
2. S.C. Wilks, W.L. Kruer, M. Tabak, A.B. Langdon, Absorption of ultra-intense laser-pulses. Phys. Rev. Lett. **69**(9), 1383–1386 (1992)
3. W. Kruer, *The Physics of Laser Plasma Interactions* (Westview Press, Boulder, 2001)
4. Eliezer, S., *The Interaction of High Power Lasers with Plasmas.* (IOP Publishing, Bristol, 2002)
5. V.L. Ginzburg, *The Propagation of Electromagnetic Waves in Plasmas* (Pergamon, New York, 1964)
6. F. Brunel, Not-so-resonant, resonant absorption. Phys. Rev. Lett. **59**(1), 52–55 (1987)
7. P. Gibbon, A.R. Bell, Collisionless absorption in sharp-edged plasmas. Phys. Rev. Lett. **68**(10), 1535–1538 (1992).
8. M.P. Kalashnikov, P.V. Nickles, T. Schlegel, M. Schnuerer, F. Billhardt, I. Will, W. Sandner, N.N. Demchenko. *Dynamics of Laser-Plasma Interaction at* $1018\,W/cm^2$. Phys. Rev. Lett. **73**(2), 260 (1994)
9. A.P. Fews, P.A. Norreys, F.N. Beg, A.R. Bell, A.E. Dangor, C.N. Danson, P. Lee, S.J. Rose. Plasma ion emission from high- intensity picosecond laser–pulse interactions with solid targets. Phys. Rev. Lett. **73**(13), 1801–1804 (1994)
10. G. Malka, J. Fuchs, F. Amiranoff, S.D. Baton, R. Gaillard, J.L. Miquel, H. Pepin, C. Rousseaux, G. Bonnaud, M. Busquet, L. Lours. Suprathermal electron generation and channel formation by an ultrarelativistic laser pulse in an underdense preformed plasma. Phys. Rev. Lett. **79**(11), 2053–2056 (1997)
11. M.H. Key, M.D. Cable, T.E. Cowan, K.G. Estabrook, B.A. Hammel, S.P. Hatchett, E.A. Henry, D.E. Hinkel, J.D. Kilkenny, J.A. Koch, W.L. Kruer, A.B. Langdon, B.F. Lasinski, R.W. Lee, B.J. MacGowan, A. MacKinnon, J.D. Moody, M.J. Moran, A.A. Offenberger, D.M. Pennington, M.D. Perry, T.J. Phillips, T.C. Sangster, M.S. Singh, M.A. Stoyer, M. Tabak, G.L. Tietbohl, M. Tsukamoto, K. Wharton, S.C. Wilks. Hot electron production and heating by hot electrons in fast ignitor research. Phys. Plasmas **5**(5), 1966–1972 (1998)
12. K.B. Wharton, S.P. Hatchett, S.C. Wilks, M.H. Key, J.D. Moody, V. Yanovsky, A.A. Offenberger, B.A. Hammel, M.D. Perry, and C. Joshi. Experimental measurements of hot electrons generated by ultraintense (> 1019 W/cm^2) laser–plasma interactions on solid-density targets. Phys. Rev. Lett. **81**(4), 822–825 (1998)
13. R.A. Snavely, M.H. Key, S.P. Hatchett, T.E. Cowan, M. Roth, T.W. Phillips, M.A. Stoyer, E.A. Henry, T.C. Sangster, M.S. Singh, S.C. Wilks, A. MacKinnon, A. Offenberger, D.M. Pennington, K. Yasuike, A.B. Langdon, B.F. Lasinski, J. Johnson, M.D. Perry, E.M. Campbell. Intense high-energy proton beams from Petawatt-Laser irradiation of solids. Phys. Rev. Lett. **85**(14), 2945 (2000)

14. S.P. Hatchett, C.G. Brown, T.E. Cowan, E.A. Henry, J.S. Johnson, M.H. Key, J.A. Koch, A.B. Langdon, B.F. Lasinski, R.W. Lee, A. J. Mackinnon, D.M. Pennington, M.D. Perry, T.W. Phillips, M. Roth, T.C. Sangster, M.S. Singh, R.A. Snavely, M.A. Stoyer, S.C. Wilks, K. Yasuike. Electron, photon, and ion beams from the relativistic interaction of Petawatt laser pulses with solid targets. Phys. Plasmas 7(5), 2076–2082 (2000)
15. F. Amiranoff. Fast electron production in ultra-short high-intensity laserplasma interaction and its consequences. Meas. Sci. Technol. 12(11), 1795–1800 (2001)
16. L. Gremillet, F. Amiranoff, S.D. Baton, J.C. Gauthier, M. Koenig, E. Martinolli, F. Pisani, G. Bonnaud, C. Lebourg, C. Rousseaux, C. Toupin, A. Antonicci, D. Batani, A. Bernardinello, T. Hall, D. Scott, P. Norreys, H. Bandulet, H. Pepin. Time-resolved observation of ultrahigh intensity laser-produced electron jets propagating through transparent solid targets. Phys. Rev. Lett. 83(24), 5015–5018 (1999)
17. M. Borghesi, A.J. Mackinnon, A.R. Bell, G. Malka, C. Vickers, O. Willi, J. R. Davies, A. Pukhov, J. Meyer-ter Vehn. Observations of collimated ionization channels in aluminum-coated glass targets irradiated by ultraintense laser pulses. Phys. Rev. Lett. 83(21) 4309–4312 (1999)
18. M. Tatarakis, J.R. Davies, P. Lee, P.A. Norreys, N.G. Kassapakis, F.N. Beg, A.R. Bell, M.G. Haines, A.E. Dangor. Plasma formation on the front and rear of plastic targets due to high-intensity laser-generated fast electrons. Phys. Rev. Lett. 81(5), 999–1002 (1998)
19. J.R. Davies, A.R. Bell, M.G. Haines, S.M. Guerin. Short-pulse high-intensity laser-generated fast electron transport into thick solid targets. Phys. Rev. E 56(6), 7193–7203 (1997)
20. A.J. Mackinnon, Y. Sentoku, P.K. Patel, D.W. Price, S. Hatchett, M.H. Key, C. Andersen, R. Snavely, R.R. Freeman. Enhancement of proton acceleration by hot-electron recirculation in thin foils irradiated by ultraintense laser pulses. Phys. Rev. Lett. 88(21), 215006 (2002)
21. Y.S. Huang, X.F. Lan, X.J. Duan, Z.X. Tan, N.Y. Wang, Y.J. Shi, X.Z. Tang, Y.X. He. Hot-electron recirculation in ultraintense laser pulse interactions with thin foils. Phys. Plasmas 14(10) (2007)
22. S.C. Wilks, A.B. Langdon, T.E. Cowan, M. Roth, M. Singh, S. Hatchett, M.H. Key, D. Pennington, A. MacKinnon, R.A. Snavely. Energetic proton generation in ultra-intense laser–solid interactions. Phys. Plasmas 8(2), 542–549 (2001)
23. P. Mora. Plasma expansion into a vacuum. Phys. Rev. Lett. 90(18), 185002 (2003)
24. T. Feurer, W. Theobald, R. Sauerbrey, I. Uschmann, D. Altenbernd, U. Teubner, P. Gibbon, E. Forster, G. Malka, J.L. Miquel. Onset of diffuse re ectivity and fast electron ux inhibition in 528-nm-laser solid interactions at ultrahigh intensity. Phys. Rev. E 56(4), 4608–4614 (1997)
25. J. Fuchs, P. Antici, E. d'Humieres, E. Lefebvre, M. Borghesi, E. Brambrink, C. A. Cecchetti, M. Kaluza, V. Malka, M. Manclossi, S. Meyroneinc, P. Mora, J. Schreiber, T. Toncian, H. Pepin, and R. Audebert. Laser-driven proton scaling laws and new paths towards energy increase. Nat. Phys. 2(1), 48–54 (2006)
26. Y. Oishi, T. Nayuki, T. Fujii, Y. Takizawa, X. Wang, T. Yamazaki, K. Nemoto, T. Kayoiji, T. Sekiya, K. Horioka, Y. Okano, Y. Hironaka, K. G. Nakamura, K. Kondo, and A. A. Andreev. Dependence on laser intensity and pulse duration in proton acceleration by irradiation of ultrashort laser pulses on a Cu foil target. Phys. Plasmas 12(7), 5 (2005)
27. J. Schreiber, F. Bell, F. Gruner, U. Schramm, M. Geissler, M. Schnurer, S. Ter-Avetisyan, B. M. Hegelich, J. Cobble, E. Brambrink, J. Fuchs, P. Audebert, D. Habs. Analytical model for ion acceleration by high-Intensity laser pulses. Phys. Rev. Lett. 97(4), 045005–4 (2006)
28. A.A. Andreev, R. Sonobe, S. Kawata, S. Miyazaki, K. Sakai, K. Miyauchi, T. Kikuchi, K. Platonov, K. Nemoto. Effect of a laser prepulse on fast ion generation in the interaction of ultra-short intense laser pulses with a limited-mass foil target. Plasma Phys. Controlled Fusion 48(11), 1605–1619 (2006)
29. H. Ruhl, S.V. Bulanov, T.E. Cowan, T.V. Liseikina, P. Nickles, F. Pegoraro, M. Roth, W. Sandner. Computer simulation of the threedimensional regime of proton acceleration in the interaction of laser radiation with a thin spherical target. Plasma Phys. Rep. 27(5), 363–371 (2001)

30. P. Mora. Thin-foil expansion into a vacuum. Phys. Rev. E **72**(5), 5 (2005)
31. A. Pukhov. Three-dimensional simulations of ion acceleration from a foil irradiated by a short-pulse laser. Phys. Rev. Lett. **86**(16), 3562–3565 (2001)
32. A. Maksimchuk, S. Gu, K. Flippo, D. Umstadter, V.Y. Bychenkov. Forward ion acceleration in thin films driven by a high-intensity laser. Phys. Rev. Lett. **84**(18), 4108–4111 (2000)
33. M. Zepf, E.L. Clark, F.N. Beg, R.J. Clarke, A.E. Dangor, A. Gopal, K. Krushelnick, P.A. Norreys, M. Tatarakis, U. Wagner, M. S. Wei. Proton acceleration from high-intensity laser interactions with thin foil targets. Phys. Rev. Lett. **90**(6), 064801 (2003)
34. E. d'Humieres, E. Lefebyre, L. Gremillet, V. Malka. Proton acceleration mechanisms in high-intensity laser interaction with thin foils. Phys. Plasmas **12**(6), 13 (2005)
35. S. Gaillard, J. Fuchs, N. Renard-Le Galloudec, T.E. Cowan. Study of saturation of CR39 nuclear track detectors at high ion uence and of associated artifact patterns. Rev. Sci. Instr. **78**(1), 13 (2007)
36. M. Hegelich, S. Karsch, G. Pretzler, D. Habs, K. Witte, W. Guenther, M. Allen, A. Blazevic, J. Fuchs, J. C. Gauthier, M. Geissel, P. Audebert, T. Cowan, M. Roth. MeV ion jets from short-pulse-laser interaction with thin foils. Phys. Rev. Lett. **89**(8), 4 (2002)
37. B.M. Hegelich, B.J. Albright, J. Cobble, K. Flippo, S. Letzring, M. Paffett, H. Ruhl, J. Schreiber, R.K. Schulze, and J. C. Fernandez. Laser acceleration of quasi-monoenergetic MeV ion beams. Nature **439**(7075), 441–444 (2006)
38. F.N. Beg, M.S.Wei, A.E. Dangor, A. Gopal, M. Tatarakis, K. Krushelnick, P. Gibbon, E.L. Clark, R.G. Evans, K.L. Lancaster, P.A. Norreys, K.W.D. Ledingham, P. McKenna, M. Zepf. Target charging effects on proton acceleration during high-intensity short-pulse lasersolid interactions. Appl. Phys. Lett. **84**(15), 2766–2768 (2004)
39. T. Toncian, M. Borghesi, J. Fuchs, E. d'Humieres, P. Antici, P. Audebert, E. Brambrink, C. A. Cecchetti, A. Pipahl, L. Romagnani, and O. Willi. Ultrafast laser-driven microlens to focus and energy-select mega-electron volt protons. Science, **312**(5772), 410–413 (2006)
40. V.F. Kovalev, K.I. Popov, V.Y. Bychenkov, W. Rozmus. Laser triggered Coulomb explosion of nanoscale symmetric targets. Physics of Plasmas, **14**(5), 053103 (2007)
41. A.P.L. Robinson, M. Zepf, S. Kar, R. G. Evans, C. Bellei. Radiation pressure acceleration of thin foils with circularly polarized laser pulses. New J. Phys. **10** (2008)
42. T. Esirkepov, M. Borghesi, S. V. Bulanov, G. Mourou, T. Tajima. Highly efficient relativistic-ion generation in the laser-piston regime. Phys. Rev. Lett. **92**(17) (2004)
43. A. Macchi, F. Cattani, T.V. Liseykina, F. Cornolti. Laser acceleration of ion bunches at the front surface of overdense plasmas. Phys. Rev. Lett. **94**(16) (2005)

Chapter 5
Laser System

The High Field Laser system at the Max-Born-Institute consists of a multi TW Ti:Sa laser synchronized to a Nd:glass laser. This system allows pump-probe experiments at intensities $\geq 10^{18}$ W/cm^2 while the maximum intensity of the Ti:Sa beam exceeds 10^{19} W/cm^2. In this chapter a short overview of the laser systems and their synchronization will be given.

5.1 Ti:Sa Laser System

The Ti:Sa laser system delivers 40 fs pulses with a pulse energy of about 1 J at a repetition rate of 10 Hz. The CPA (Sect. 2.1) system consists of a commercial oscillator (femto source) and three multi-pass amplifiers pumped by frequency-doubled Nd:YAG lasers [1]. A schematic picture of the system is shown in Fig. 5.1.

For experiments using high power lasers systems the temporal contrast of the laser pulse is highly important. Reaching intensities of about 10^{19} W/cm^2 the contrast has to be better than 10^6 to avoid a remarkable influence of ionization by pre-pulses or Amplified Spontaneous Emission (ASE). The main amplification is usually done by the first amplifier with an amplification factor of about 10^6. Hence this stage is crucial for the final pulse contrast. In difference to multi-pass amplifiers, regenerative amplifiers implement pre-pulses by the out-coupling of the amplified pulse due to a Pockels cell. Nevertheless the ASE pedestal creates a nanosecond pedestal in front of the main peak. To reduce this pedestal a fast Pockels cell was installed after the first amplifier. In Fig. 5.2 a THG-autocorrelator measurement is shown.

The laser system delivers pulses with a contrast between 10^7 and 10^8 (dependent on the pump conditions of the Ti:Sa crystals and the timing of the fast Pockels

T. Sokollik, *Investigations of Field Dynamics in Laser Plasmas with Proton Imaging,* Springer Theses, DOI: 10.1007/978-3-642-15040-1_5,
© Springer-Verlag Berlin Heidelberg 2011

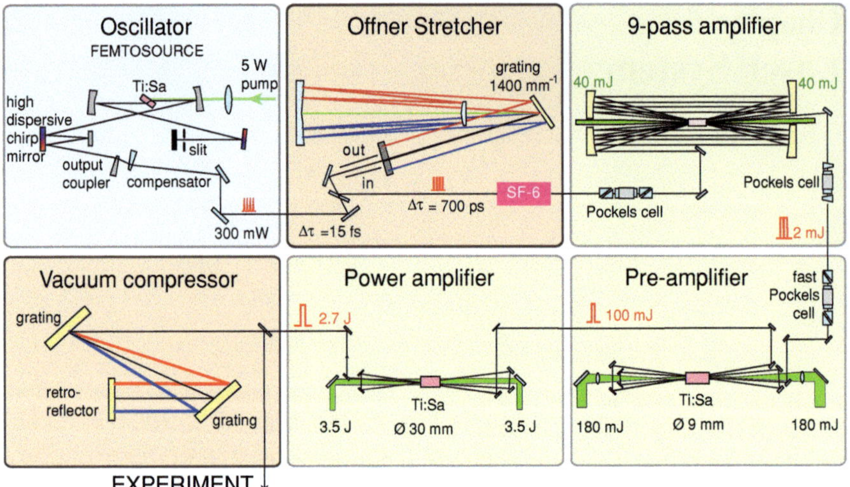

Fig. 5.1 Schematic setup of the MBI TW Ti:Sa laser

Fig. 5.2 Contrast measurement of the laser pulse done with a third-order autocorrelator

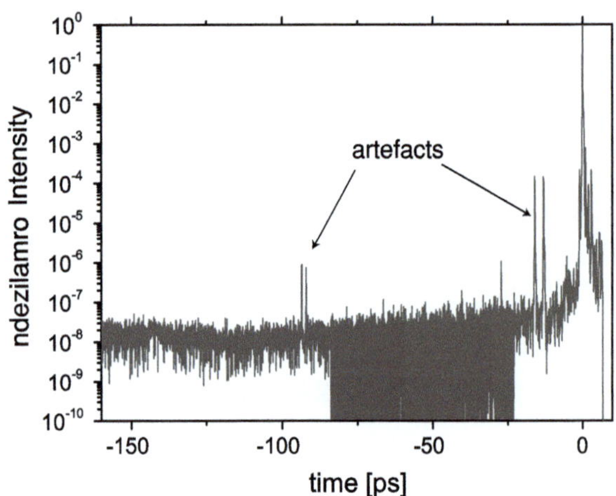

cells) related to the main peak which are well suited for experiments with high intensities e.g., the ion acceleration.

The high energy and the short duration of the pulses require a large beam diameter to avoid damage of the optical components. Thus, the beam diameter after the compression is about 60 mm. In the experiments the beam is focused by off-axis parabolic mirrors to a focal spot of a few micrometer diameter. To reach such small foci and thus high intensities a flattening of the beam wave front is required, which is distorted after propagating through the amplifier [2]. Thus, a deformable mirror and a Shack-Hartmann wave front sensor are installed after the compressor.

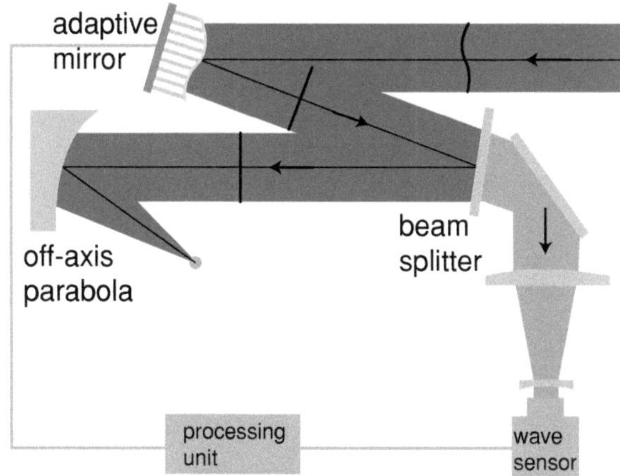

Fig. 5.3 Implementation of the adaptive mirror in the laser beam line. The pulse is split to measure the wave front by the Shack-Hartmann wave sensor. The processing unit calculates the necessary deformation of the mirror to flatten the wave front

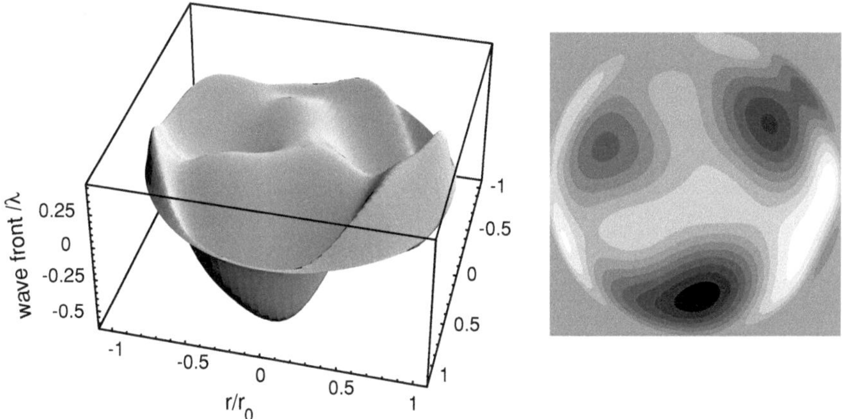

Fig. 5.4 The reconstructed wave front, calculated with Zernike polynomials delivered by a measurement with the Shack-Hartmann wave front sensor

The Shack-Hartmann sensor consist of a micro-lens array. The resulting foci are registered by a CCD camera. Small deviations of the focus positions are used to reconstruct the wave front [2]. From the reconstructed wave front a computer is calculating the best surface deformation of the mirror for compensating the wave front distortion, which is then realized by piezo elements behind the mirror surface.

Zernike polynomials (Z_n^m) are used to describe the wave front mathematically. They are products of angular functions and radial polynomials which are developed from Jacobi polynomials [3]. The first orders of the polynomials represent simple structures e.g. tilts, astigmatism, coma and defocus. In Appendix A a table of polynomials and the mathematical description is given. In Fig. 5.4 a reconstructed wave front with a RMS of 0.07 λ is shown.

To focus the beam an off-axis parabolic mirror (f/1.5) with a focal length of 15 cm is used in the experiments. In Fig. 5.5 the focus distribution is shown. The diffraction rings are a result of the focus imaging with an additional lens. The gaussian distribution has a FWHM of 5 μm and about 28% of the energy is inside the first order of diffraction.

5.2 Nd:glass Laser System

The active medium in the amplifiers of the glass laser are glass rods doped with Neodymium. Thus, the laser wave length is about 1053 nm. The spectral bandwidth (~ 5 nm) of the amplified pulses allows temporal pulse durations between 1 and 2 ps [4, 5]. The laser system uses the CPA-technique [6] to amplify the laser pulses up to 10 J. The amplification takes place in a chain of amplifiers with growing diameter, pumped by flash lamps. Between the amplifiers spatial filters are installed to clean the beam profile and to adapt the beam diameter. Due to the large-sized glass rods in the amplifiers and their low thermal conductivity the thermal balancing by a water cooling system needs several minutes. Hence the repetition rate is limited and the full laser system delivers one shot per 15 min. In Fig. 5.6 the schematic setup of the system is shown.

The pulse duration was measured with a single-shot second-order autocorrelator. The beam is split into two beams which are then overlapped at a defined angle in a nonlinear crystal.

The Second Harmonic (SH) signal is generated proportional to the product of the intensities of the two beams and is recorded by a CCD camera. From the width of the SH signal the pulse duration can be determined assuming the temporal shape of the pulse (e.g., gaussian). In Fig. 5.7 the autocorrelation function is plotted with a gaussian fit. The pulse duration was estimated to be about 2 ps.

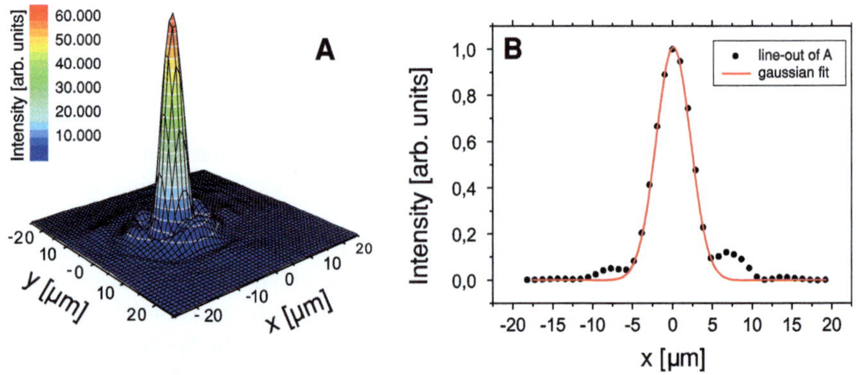

Fig. 5.5 Measured focus distribution with a 5 μm FWHM

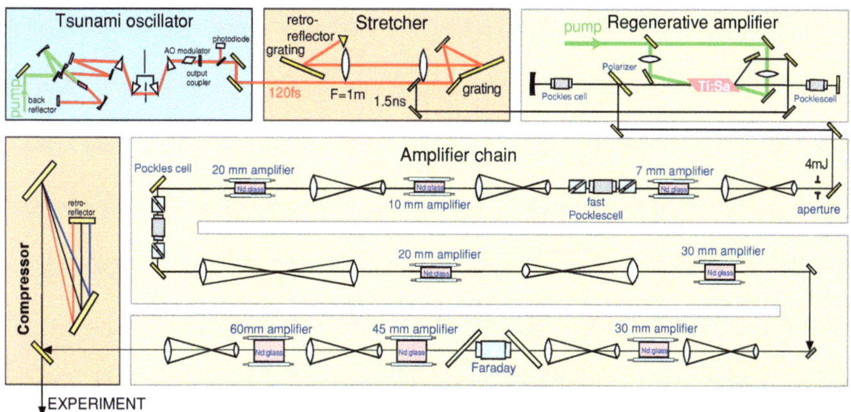

Fig. 5.6 Schematic setup of the Nd:glass laser

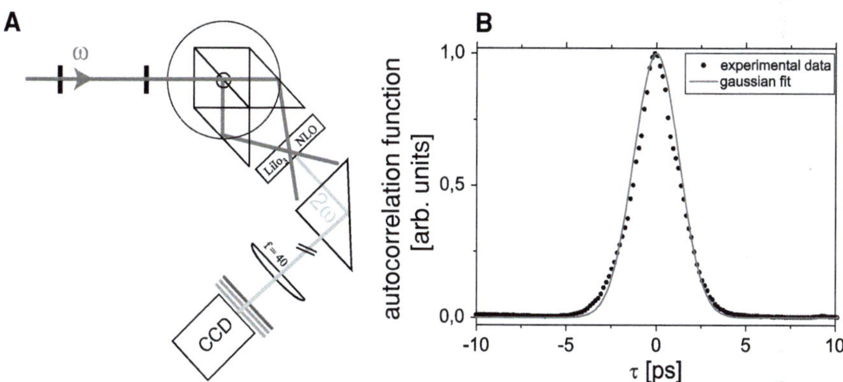

Fig. 5.7 a Setup of the second-order autocorrelator. **b** Measured autocorrelator function with an estimated pulse duration of 2 ps

A pulse contrast measurement was done by recording two photodiode signals of one laser pulse with different neutral density filters (differ by a factor of 10^{-3}). Thus, the main peak and the pre-pulses could be measured (Fig. 5.8).

With this measurement the contrast (ratio of the main peak to the pre-pulses) was estimated to be about $8 \cdot 10^6$.

5.3 Synchronization

The two laser systems, the Ti:Sa and the Nd:glass laser are synchronized to each other. For this the oscillators are locked electronically to a quartz clock defining

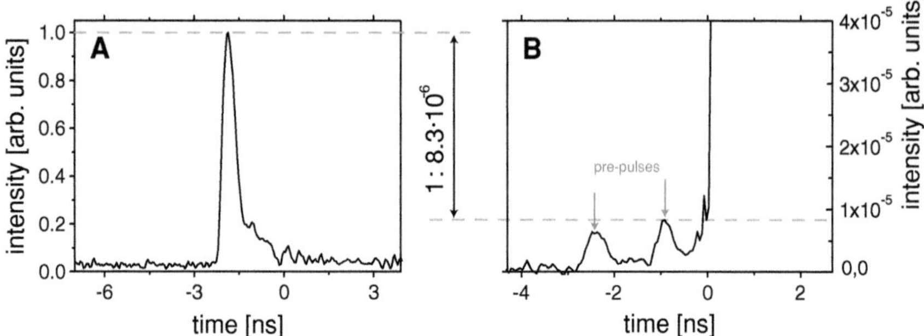

Fig. 5.8 Contrast measurement of the laser pulse using two photodiodes. **a** In front of one of the diodes an additional neutral density filter with a transmission of 10^{-3} was placed. **b** The other diode is saturated but delivers a reasonable signal before the main peak of the pulse arises. Thus, the contrast between the pre-pulses and the main peak can be estimated about $8 \cdot 10^{6}$

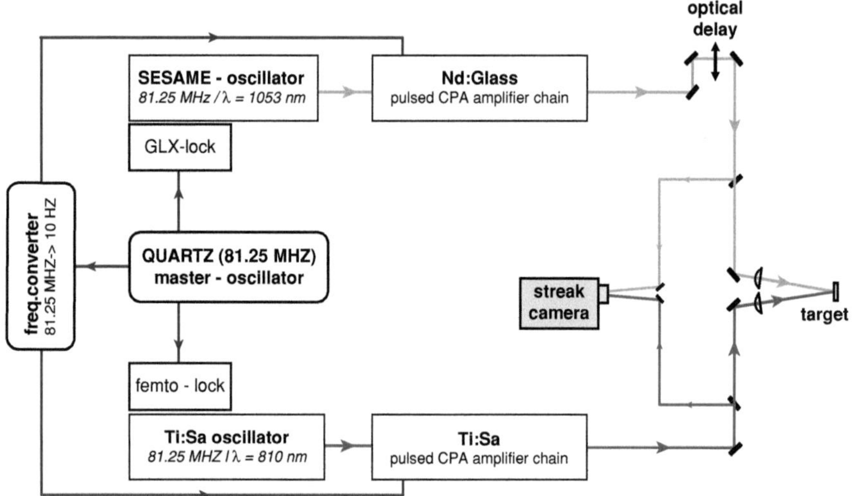

Fig. 5.9 Synchronization scheme of the two laser systems

the repetition rate and the phase. A frequency of 81.25 MHz is achieved in both oscillators by varying the length of the resonators. The synchronization scheme is shown in Fig. 5.9.

The electronic locking allows an overlap of the pulses with a precision of a few picoseconds. The pulses arrive at target-chamber in a temporal window of 3 ns. An optical delay stage was used to compensate the residual delay. The temporal jitter of the two laser pulses was measured with a streak camera (Hamamatsu C1587). Such a device transforms the light pulse into an electron bunch which is deflected by an electric field varying in time. Hence the temporal structure can be measured and also the time delay between two laser pulses can be determined. The resolution of the used camera was about 2 ps.

In Fig. 5.10 two measurements with the streak camera are shown. Between the two shots (with maximum laser energy) the relative temporal distance between Nd:glass laser pulse and the Ti:Sa laser pulse changes by about 6 ps. This represents the averaged jitter, fluctuating from shot to shot. The measured jitter using the frontend of both lasers was about 3 ps. The difference can be explained by the influence of the amplifiers on the electronic equipment. Nevertheless, it was more than sufficient for the applied experiments.

To align the optical delay stage in the 10 Hz operation mode, the plasma shadowgraphy can be used as a simple method. The Ti:Sa laser is focused inside the experimental chamber. The intense laser radiation creates a plasma in air which acts like a concave lens due to the locally increase of the electron density (see Chap. 3). The Nd:glass laser beam is influenced by this plasma lens and a shadow of the plasma spot can be observed (Fig. 5.11a). By varying the temporal

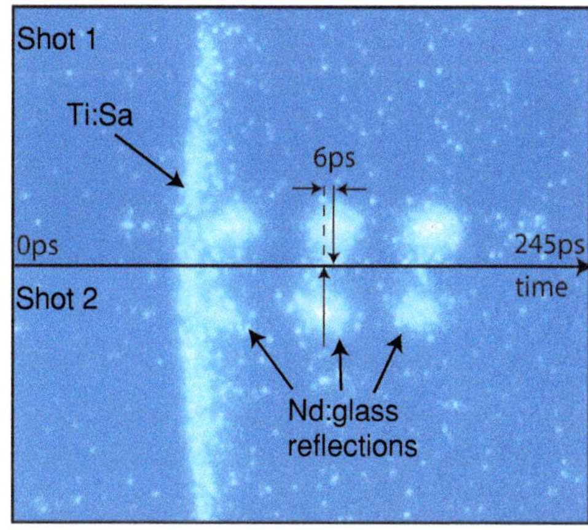

Fig. 5.10 Two different measurements with the streak camera represent the estimated jitter between the Ti:Sa and the Nd:Glass laser pulse of about 6 ps

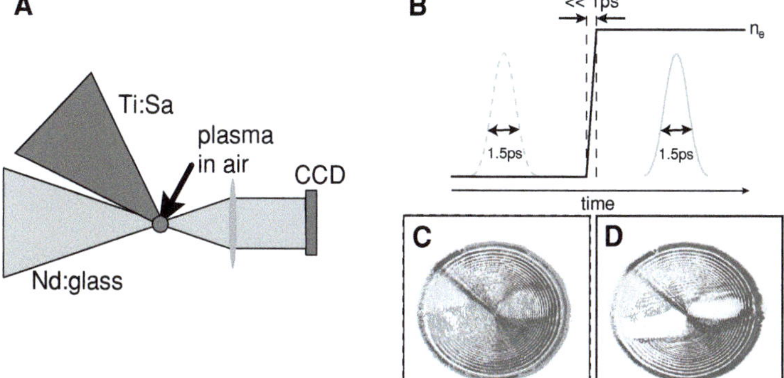

Fig. 5.11 a Setup of the synchronization using the shadowgraphy method. Both beams are focused in air. **b** The measured Nd:glass laser signal shows a very sensitive behavior either if the glass laser pulse is temporally before the Ti:Sa pulse or contrariwise. If the Ti:Sa creates the plasma before the Nd:glass laser arrives the beam is partly influenced by the plasma

delay the position can be found where the laser beam is barely distorted. At this position both laser pulses are temporally overlapping in the plasma (cf. Fig. 5.11b–d). The precision of this method is in the range of the Nd:glass laser pulse duration.

The foci of the two laser beams have not to be overlapping necessarily. Even in the unfocused or divergent beam the influence of the plasma and thus the influence of the other laser is visible. If the Ti:Sa laser is used to probe the created plasma no CCD camera is needed. The influence can be directly observed on a IR sensor card. This method can be used to temporally overlap laser pulses even of different wavelengths and in a broad temporal window. Since the plasma lifetime is relatively long (nanoseconds), the position where the plasma is created can be found easily. This is an advantage compared to other correlation devices like streak camera or cross-correlators which have usually a small temporal observation window of a few picoseconds only.

References

1. M.P. Kalachnikov, V. Karpov, H. Schonnagel, W. Sandner, 100- Terawatt titanium–sapphire laser system. Laser Phys. **12(2)**, 368–374 (2002)
2. Z.H. Wang, Z. Jin, J. Zheng, P. Wang, Z. Y. Wei, J. Zhang, Wavefront correction of high-intensity fs laser beams by using closed-loop adaptive optics system. Sci. China Ser. G Phys. Mech. Astron. **48(1)**, 122–128 (2005)
3. R.J. Noll, Zernike polynomials and atmospheric-turbulence. J. Opt. Soc. Am. **66(3)**, 207–211 (1976)
4. G. Priebe, K.A. Janulewicz, V.I. Redkorechev, J. Tummler, P.V. Nickles, Pulse shape measurement by a non-collinear third-order correlation technique. Opt. Commun. **259(2)**, 848–851 (2006)
5. F. Billhardt, M. Kalashnikov, P.V. Nickles, I. Will, A high-contrast Ps-terawatt Nd-glass laser system with fiberless chirped pulse amplification. Opt. Commun. **98(1–3)**, 99–104 (1993)
6. D. Strickland, G. Mourou, Compression of amplified chirped optical pulses. Opt. Commun. **56(3)**, 219–221 (1985)

Part II
Proton Beam Characterization

Chapter 6
Proton and Ion Spectra

Proton and ion beams accelerated indirectly by laser pulses exhibit very special attributes, determined by the acceleration mechanism and the target geometry. In the following chapter the characteristics of the proton and ion spectra will be discussed for different target types and laser parameters. A short overview of usual characteristics, but also of observed irregularities in the spectra will be given. This is followed by Chaps. 6 and 7 where the beam emittance and the proton source dynamics will be discussed.

6.1 Thomson Spectrometer

The proton and ion spectra can be measured with a Thomson spectrometer. The setup is shown in Fig. 6.1. It consists of a pinhole and a magnet combined with electric field plates. The small ion beamlet propagating in z-direction through the pinhole is deflected due to the magnetic field B_y:

$$x = \frac{Q B_y l L}{m_i v} \tag{6.1}$$

and the different ion species are separated by the additional deflection by the electric field:

$$y = \frac{Q E_y l L}{m_i v^2}. \tag{6.2}$$

Substituting v^2 by Eq. 6.1 the deflection curve at the detector can be described by a parabolic function:

$$y = \frac{m_i E_y}{Q B_y^2 l L} \cdot x^2, \tag{6.3}$$

T. Sokollik, *Investigations of Field Dynamics in Laser Plasmas with Proton Imaging,* Springer Theses, DOI: 10.1007/978-3-642-15040-1_6,
© Springer-Verlag Berlin Heidelberg 2011

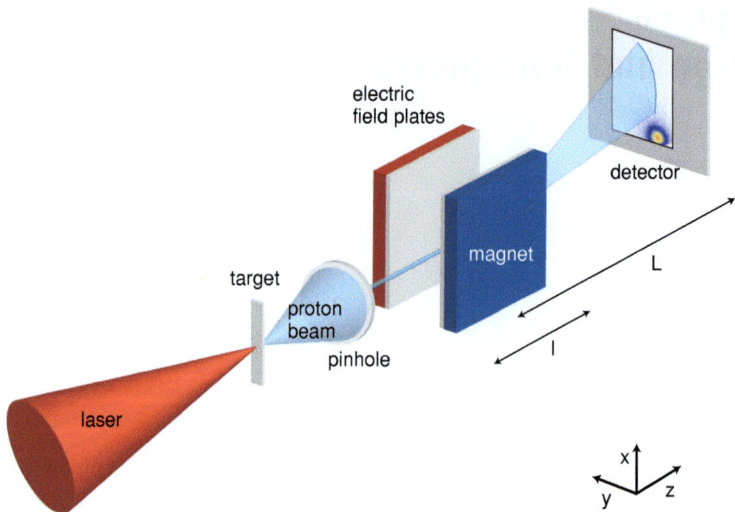

Fig. 6.1 Setup of a Thomson spectrometer for laser accelerated ion beams

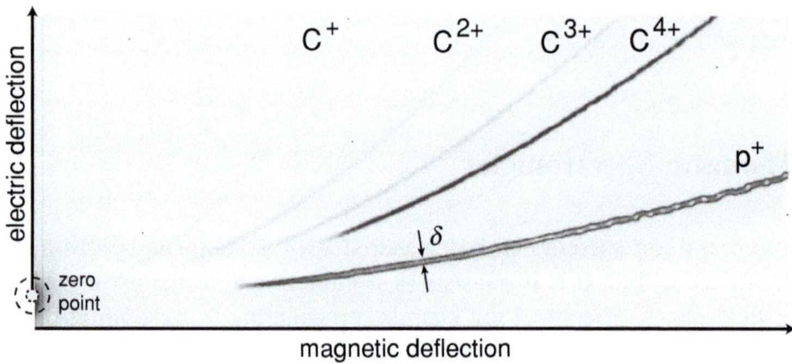

Fig. 6.2 Image, taken with a CCD camera, of the illuminated phosphor screen coupled to the MCP. Five traces are clearly visible representing different ion species, protons and positively charged carbon ions. The zero point determines the projection of the pinhole without any deflection

where E_y and B_y are the electric and magnetic field, Q the charge of the ion and m_i its mass. l is the effective extension of the electric and magnetic field and L the distance between magnet and detector. Equations (6.1–6.3) are only valid for small angle of deflections. In Fig. 6.2 a typical picture of the ion spectra is shown. The particles are detected with a multi-channel plate (MCP) coupled to a phosphor screen and an attached CCD camera.

Apart from protons also carbon ions are detected. They are accelerated from the contamination layer on the rear side of the target - similar to the protons. The contamination layer consists of hydrocarbon and water mainly [1, 2]. The shown spectra were generated by irradiating a 5 μm thick Titanium foil with the Ti:Sa laser at an intensity of about 10^{19} W/cm^2. In Fig. 6.3 the traces have been integrated and the counts were calibrated with the detection sensitivity of the MCP [3]. The proton cut-off energy in this measurement is about 2 MeV whereas energies up to 4–5 MeV were observed under similar conditions.

The energy resolution of the Thomson spectrometer is defined by the width of the parabolic trace δ [4] (cf. Fig. 6.2). In the presented setup the energy resolution ΔE can be estimated as follows:

$$\frac{\Delta E}{E} \approx \frac{\delta}{x}, \tag{6.4}$$

where x is the deflection caused by the magnetic field (Eq. 6.1) and δ is defined by the spectrometer setup:

$$\Delta E \approx \frac{\delta}{a} E^{3/2}, \quad a = \frac{QBlL}{\sqrt{2m}}. \tag{6.5}$$

The energy resolution for the spectrum in Fig. 6.3 is plotted in Fig. 6.4 with a δ of about 0.42 mm estimated from the MCP image (Fig. 6.2). The proton spectra are usually continuous up to a sharp cut-off representing the highest proton energies.

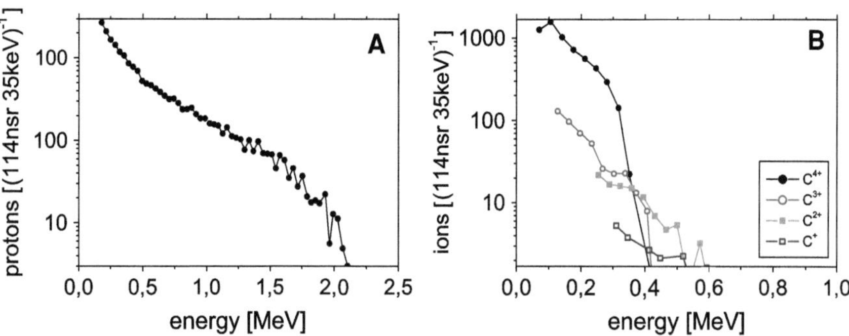

Fig. 6.3 a Proton spectra extracted from Fig. 6.2. **b** Spectra of the carbon ions

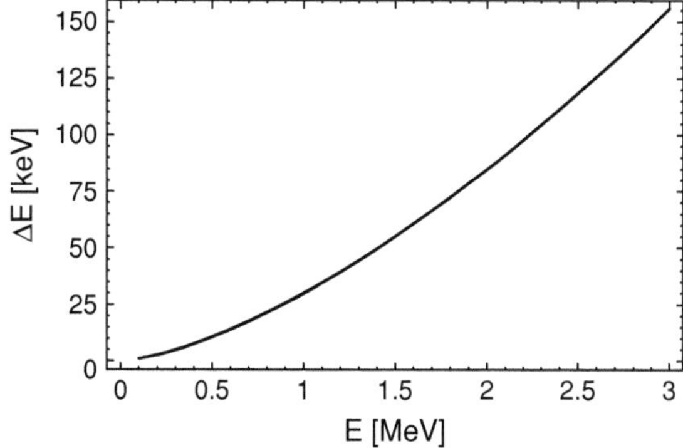

Fig. 6.4 Energy resolution calculated for the spectra measurement of Fig. 6.2 dependent on the proton energy

However, for specific target types ion spectra with dips [5, 6] or quasi-monoenergetic [7–10] features have been observed. Since the generation of monoenergetic ion beams is of high interest for many applications, it will be discussed shortly in the next section.

6.2 Quasi-Monoenergetic Deuteron Bursts

Irradiating water or heavy water droplets with about 20 μm diameter non-continuous deuteron or proton spectra are observed. Spectral dips [5, 6] or even quasi-monoenergetic features [7, 8] were seen in the measurements. This behavior is fluctuating from shot to shot since it highly depends on laser parameters [8].

Due to the generation process of the droplets (Sect. 11.2) no contamination layer is covering the surface. The detected Oxygen ions (Fig. 6.5) are accelerated from the first ionized atom layers of the droplet surface. The analyzed spectra are plotted in Fig. 6.6 showing a monoenergetic deuteron beam around 2 MeV with a

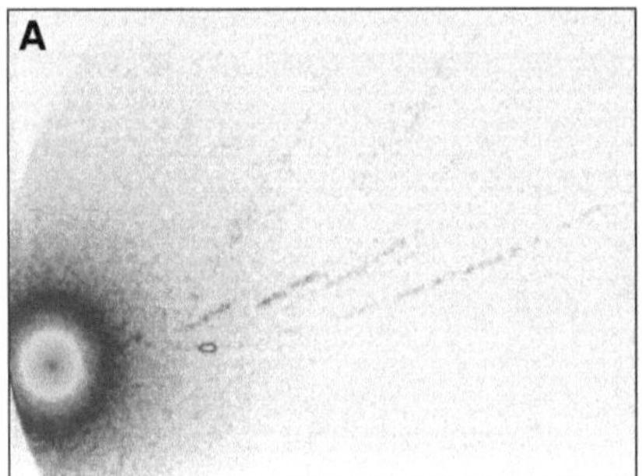
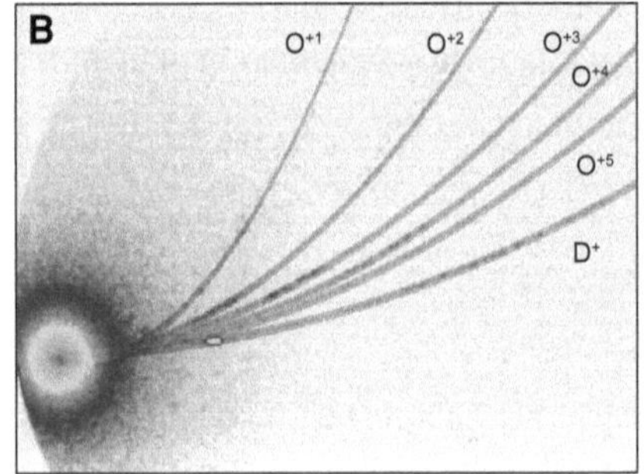

Fig. 6.5 **a** Image of the Deuteron an Oxygen spectra emitted from an illuminated heavy water droplet. **b** The ion Thomson parabolas are drawn to guide the eyes and to visualize the ion traces

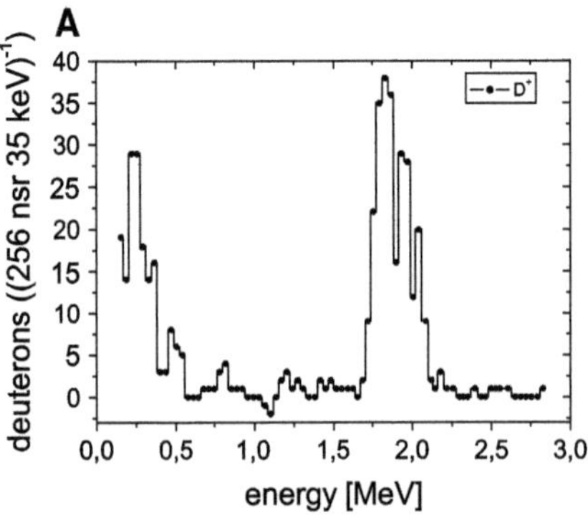
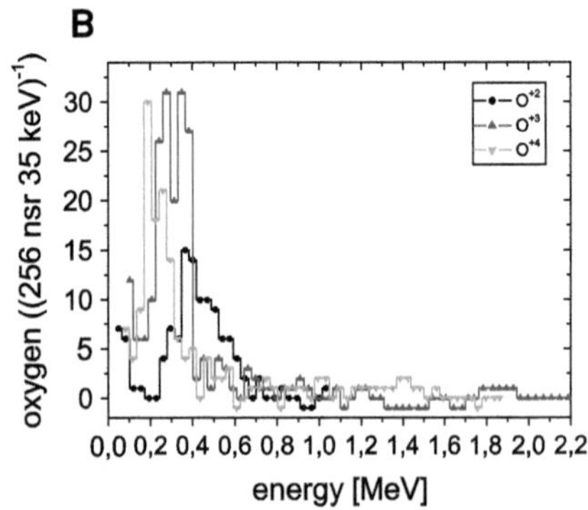

Fig. 6.6 **a** The Deuteron spectrum of Fig. 6.5 shows a narrow energy bandwidth of about 300 keV. **b** The Oxygen spectra from the same measurement

bandwidth of 300 keV. To explain the formation of such a monoenergetic feature, a model related to a spatial separation of two ion species can be used. More details about the experiment and the interpretation can be found in reference [7, 8]. The monoenergetic feature was only observed in laser forward direction. This can be an indication for hot electron currents propagating around the droplet leading to an enhancement of the electrostatic field at the rear side. Further investigations of this issue were done by proton imaging experiments and PIC-simulation in Chap. 11 delivering a strong evidence for this theses.

6.3 Irregularities of the Thomson Parabolas

Looking closer to the measured spectra deviations from the parabolic shape can be observed. For a detailed investigation a magnifying Thomson spectrometer (12-fold) was used similar to reference [11, 12]. In Fig. 6.7 such magnified spectra are shown.

At lower proton energies fluctuations ("wiggly" structures) in the parabola appear. They are caused by fluctuation of the proton beam pointing [11] (see also Chap. 8 and Sect. 12.2). In some records the proton spectrum split into two parallel traces. This can be explained by two spatially separated proton sources [12, 13]. They result from two different absorption mechanisms. The laser pulse drives the electrons ponderomotively or by resonance absorption. Both mechanisms couple energy into electrons which transfer the energy to the ions. Which one of these two absorption mechanisms dominates, depends on the plasma scale length created by the ASE level. Therefore, the dominating mechanism can be chosen by manipulating the contrast of the laser pulse. These two different electron populations could be identified by measuring Ĉerenkov emission from the rear side of the target.

Therefore a 50 μm tesa foil was attached at the rear side of a 6 μm thick aluminium target which was irradiated by a 40 fs laser pulse with an intensity of $2 \cdot 10^{19}$ W/cm^2. If the electron velocity is higher than the light velocity in a medium

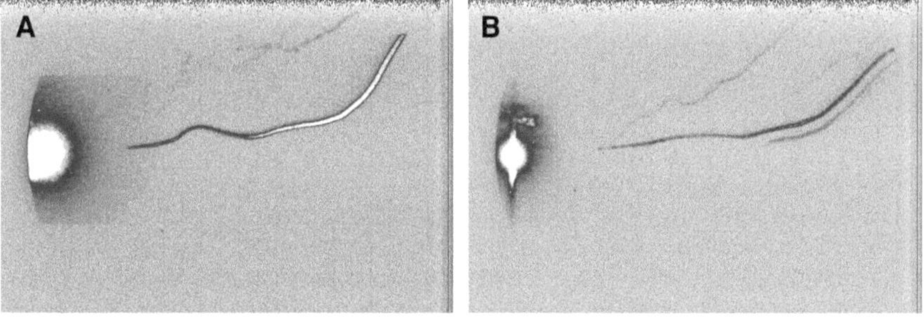

Fig. 6.7 **a** Magnified proton image showing variances from the Thomson parabola. **b** Apart from the fluctuation a second proton trace parallel to the first one appears

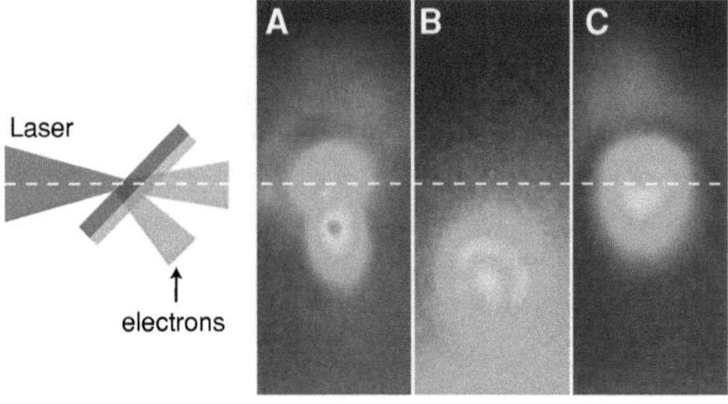

Fig. 6.8 The laser irradiates the target at 45°. The Ĉerenkov medium attached at the rear surface of the aluminium target creates the Ĉerenkov light when energetic electron currents enter the medium

Ĉerenkov light is emitted [14]. By detecting the generated light with a CCD camera, information of the electron currents can be gained. In Fig. 6.8 three different measurements are shown. Two different electron currents have been identified, generated either by ponderomotive acceleration or by resonance absorption. The laser pulse parameters define the dominant process although both mechanisms can occur simultaneously. Both electron currents lead to an ion acceleration and can be connected with spatially separated ion sources (cf. Fig. 6.7b). In reference [12] a more detailed discussion of this experiment can be found.

References

1. M. Hegelich, S. Karsch, G. Pretzler, D. Habs, K. Witte, W. Guenther, M. Allen, A. Blazevic, J. Fuchs, J.C. Gauthier, M. Geissel, P. Audebert, T. Cowan, M. Roth, MeV ion jets from short-pulse-laser interaction with thin foils. Phys. Rev. Lett. **89**(8), 4 (2002)
2. S.J. Gitomer, R.D. Jones, F. Begay, A.W. Ehler, J.F. Kephart, R. Kristal, Fast ions and hot-electrons in the laser-plasma interaction. Phys. Fluids **29**(8), 2679–2688 (1986)
3. S. Ter-Avetisyan, M. Schnurer, P.V. Nickles, Time resolved corpuscular diagnostics of plasmas produced with high-intensity femtosecond laser pulses. J. Phys. D Appl. Phys. **38**(6), 863–867 (2005)
4. R.F. Schneider, C.M. Luo, M.J. Rhee, Resolution of the Thomson spectrometer. J. Appl. Phys. **57**(1), 1–5 (1985)
5. S. Busch, M. Schnurer, M. Kalashnikov, H. Schonnagel, H. Stiel, P.V. Nickles, W. Sandner, S. Ter-Avetisyan, V. Karpov, U. Vogt, Ion acceleration with ultrafast lasers. Appl. Phys. Lett. **82**(19), 3354–3356 (2003)
6. S. Ter-Avetisyan, M. Schnurer, S. Busch, E. Risse, P.V. Nickles, W. Sandner, Spectral dips in ion emission emerging from ultrashort laser-driven plasmas. Phys. Rev. Lett. **93**(15), 4 (2004)
7. S. Ter-Avetisyan, M. Schnurer, P.V. Nickles, M. Kalashnikov, E. Risse, T. Sokollik, W. Sandner, A. Andreev, V. Tikhonchuk, Quasimonoenergetic deuteron bursts produced by ultraintense laser pulses. Phys. Rev. Lett. **96**(14), 145006 (2006)
8. A.V. Brantov, V.T. Tikhonchuk, O. Klimo, D.V. Romanov, S. Ter- Avetisyan, M. Schnurer, T. Sokollik, P.V. Nickles, Quasi-monoenergetic ion acceleration from a homogeneous composite target by an intense laser pulse. Phys. Plasmas **13**(12), 10 (2006)
9. B.M. Hegelich, B.J. Albright, J. Cobble, K. Flippo, S. Letzring, M. Paffett, H. Ruhl, J. Schreiber, R.K. Schulze, J.C. Fernandez, Laser acceleration of quasi-monoenergetic MeV ion beams. Nature **439**(7075), 441–444 (2006)

10. H. Schwoerer, S. Pfotenhauer, O. Jackel, K.U. Amthor, B. Liesfeld, W. Ziegler, R. Sauerbrey, K.W.D. Ledingham, T. Esirkepov, Laserplasma acceleration of quasi-monoenergetic protons from microstructured targets. Nature **439**(7075), 445–448 (2006)
11. J. Schreiber, S. Ter Avetisyan, E. Risse, M.P. Kalachnikov, P.V. Nickles, W. Sandner, U. Schramm, D. Habs, J. Witte, M. Schnurer, Pointing of laser-accelerated proton beams. Phys. Plasmas **13**(3), 033111 (2006)
12. S. Ter-Avetisyan, M. Schnurer, T. Sokollik, P.V. Nickles, W. Sandner, H.R. Reiss, J. Stein, D. Habs, T. Nakamura, K. Mima, Proton acceleration in the electrostatic sheaths of hot electrons governed by strongly relativistic laser-absorption processes. Phys. Rev. E **77**(1), 016403 (2008)
13. P. McKenna, D.C. Carroll, R.J. Clarke, R.G. Evans, K.W.D. Ledingham, F. Lindau, O. Lundh, T. McCanny, D. Neely, A.P.L. Robinson, L. Robson, P.T. Simpson, C.G. Wahlstrom, M. Zepf, Lateral Electron Transport in High-Intensity Laser-Irradiated Foils Diagnosed by Ion Emission. Phys. Rev. Lett. **98**(14), 145001 (2007)
14. J. Stein, E. Fill, D. Habs, G. Pretzler, K. Witte, Hot electron diagnostics using X-rays and Cerenkov radiation. Laser Part. Beams **22**(3), 315–321 (2004)

Chapter 7
Beam Emittance

One important attribute of laser accelerated proton beams is the beam emittance which is a measure of the laminarity of the beam. A low emittance is essential for all kinds of proton imaging experiments. A distinction is drawn between transverse and longitudinal emittance. The transversal emittance delivers a value for the spread in angle, relative to the axes of propagation, caused by random components of transverse velocities [1]. On the other hand the longitudinal emittance considers the "velocity chirp" [2] of the proton beam and is defined by the energy-time product of the beam envelope. Although the energy spread of laser accelerated proton beams is large (e.g. 0–10 MeV), the short acceleration time (at least <10 ps) leads to a longitudinal emittance of less than 10^{-4} eV s. In [2] the longitudinal emittance was estimated by PIC-simulations being $< 10^{-7}$ eV s. This is several orders of magnitudes smaller compared to typical values of 0.5 eVs (CERN SPS) [3] for conventional accelerators. Thus, laser accelerated proton beams can be used for time resolved measurements. The longitudinal emittance delivers the upper limit for the time resolution (cf. Sect. 9.3).

As the longitudinal emittance represents the limit of the temporal resolution, the transversal emittance delivers the limit of the spatial resolution in the proton imaging experiments. Compared to conventional proton accelerators, laser accelerated protons have a much smaller (1–2 orders of magnitude) transversal emittance. For a proton beam propagating in z-direction, the transverse emittance ε_x is determined by [1]:

$$\varepsilon_x = \frac{1}{\pi} \int \int dx\, dx', \qquad x' = \frac{dx}{dz} = \frac{v_x}{v_z}. \tag{7.1}$$

The space defined by the coordinates $(x,\ x',\ y,\ y')$ is called trace space as it describes the trace of the particles. The emittance is the volume or area occupied by a distribution in the trace space [1]. Often this volume is divided by π—whereas the symbol π is appended to the units (π mm mrad).

T. Sokollik, *Investigations of Field Dynamics in Laser Plasmas with Proton Imaging*, Springer Theses, DOI: 10.1007/978-3-642-15040-1_7,
© Springer-Verlag Berlin Heidelberg 2011

For conventional accelerators it is useful to define an emittance which is independent on further acceleration (in z-direction), to get an indication of degeneration of the beam quality. Thus, the normalized emittance was introduced [1]. The normalized emittance is invariant when the beam is accelerated. The volume in $x - p_x$ space is integrated and divided by $\pi\, m_0\, c$:

$$p_x = x'\,(\beta\gamma)\,(m_0\, c), \tag{7.2}$$

$$\varepsilon_{nx,r} = (\beta\gamma)\,\varepsilon_x. \tag{7.3}$$

For non-relativistic beams the volume in $x - v_x$ space is integrated and divided by $\pi\, c$:

$$\varepsilon_{nx} = \beta\,\varepsilon_x \approx \frac{v_z}{c}\,\varepsilon_x. \tag{7.4}$$

In the following, experiments similar to reference [4] have been applied to estimate the emittance for proton beams generated with the TW Ti:Sa laser at the Max-Born-Institute.

7.1 Virtual Source

The model of the virtual proton source [4] describes the propagation of the proton beam after the acceleration process, when the beam is not interacting with itself anymore (e.g. space charge effects in non-neutral and dense ion beams are negligible). Thus, the proton trajectories are straight lines and can be traced back to a virtual source in front of the target.

In Fig. 7.1 the model is visualized. Protons are accelerated from the rear side of a laser irradiated target from an area with the lateral extension ρ and with the emission angle Θ. The proton beam is detected with a MCP (see Sect. 9.2) at the distance L from the target. In front of the target, at the distance x_v the virtual source is located with the source size a_v. The fluctuation of the emission angles

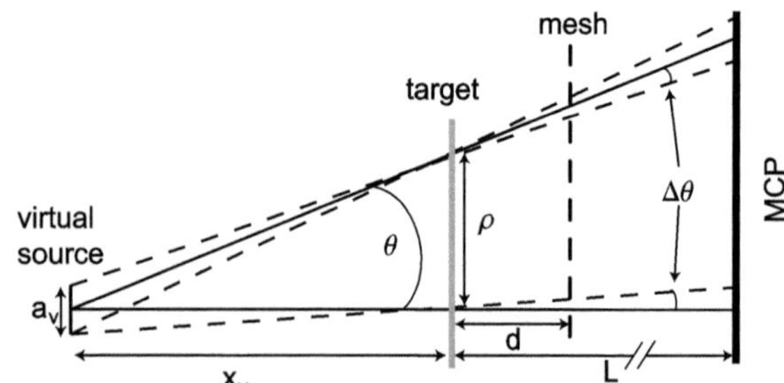

Fig. 7.1 Protons are emitted from an area with the diameter ρ. The elongated proton trajectories cross in front of the target in the virtual source. The virtual source size a_v and the distance of the virtual source x_v define the angular spread $\Delta\Theta$

$\Delta\Theta$ can be estimated by geometrical considerations (cf. Eq. 7.7). Thus, the size of the virtual source and its distance determine the degree of laminarity.

For a perfectly laminar beam the virtual source size would be almost zero. The distance x_v between the virtual source and the real source size defines the emission angle and has to be taken into account for the determination of the magnification (M) in imaging experiments (Chap. 9). For lower magnifications (<10) M can be estimated by $M = L/d$, for higher magnifications the distance of the virtual source cannot be neglected anymore:

$$M = \frac{L + x_v}{d + x_v}.\tag{7.5}$$

7.2 Measurement of the Beam Emittance

The normalized transverse emittance can be determined by Eq. 7.4 and using $\varepsilon_x \approx x_0\, x_0' = \rho\, \Delta\Theta$ [1–4]:

$$\varepsilon_{nx} \approx \frac{v_z}{c} \cdot \rho \cdot \Delta\Theta.\tag{7.6}$$

In Fig. 7.1 the setup is shown which was used to estimate the beam emittance. A mesh was inserted into the proton beam—intersecting the beam in small beamlets. From the detected mesh image the angular spread ($\Delta\Theta = x_0'$) can be determined [4] by:

$$\Delta\Theta \approx \frac{a_v}{2x_v}.\tag{7.7}$$

The distance of the virtual source x_v is defined by the magnification M of the mesh structure and Eq. 7.5. The virtual source size a_v was estimated by a fit-calculation simulating the particle traces with an inserted mesh (scattering was neglected). A larger virtual source size leads to a more blurred mesh structure in the image since the angular spread increases. Varying the virtual source size in the simulation, the shape of the imaged mesh can be fitted to the experimental results.

A scan through the imaged picture was made by integrating the measured proton numbers within a narrow stripe (indicated in Fig. 7.2b). This profile is plotted in Fig. 7.2a together with the result of a simulation with a virtual source size of $8\,\mu$m which delivers the best fit.

Now the the angular spread $\Delta\Theta$ can be estimated using Eq. 7.7. To get the normalized transverse emittance finally, the real source size ρ is needed. It was estimated by measuring the whole proton beam. Therefore a CR39 plate [5] was placed in a short distance after the mesh. In Fig. 7.3b an image of the etched plate is shown. The source size ρ was estimated roughly to be about $80\,\mu$m for protons between 0.8 and 1.2 MeV. One has to note that the accuracy of this estimation is

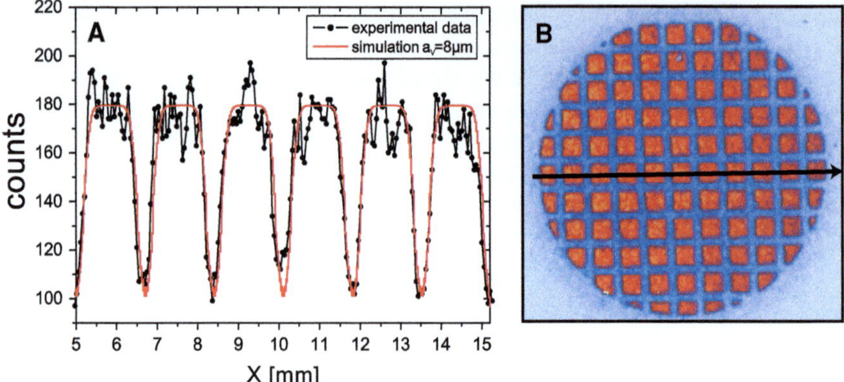

Fig. 7.2 **a** Profile of the proton image of a mesh (**b**) including a simulation where a virtual source size of $a_v = 8\,\mu$m was assumed. **b** Magnified mesh image (M $\approx$ 34-fold). The black line indicates the readout of the profile shown in (**a**)

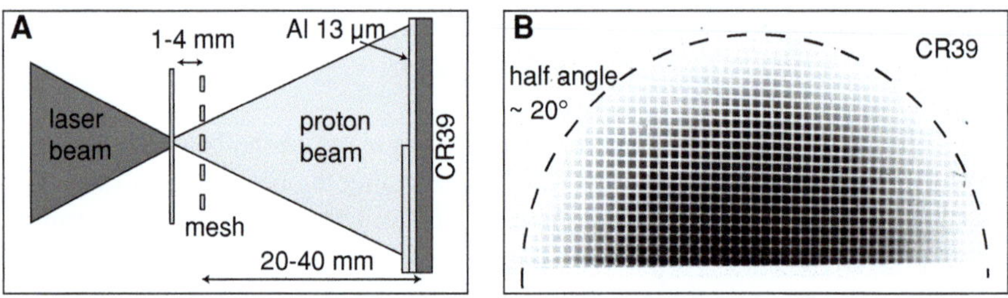

Fig. 7.3 **a** Setup with a CR39 plate for proton detection. The detector is placed close to the proton source to measure the whole beam. Two aluminium foils are attached in front of the CR39 plate to estimate the proton energy interval. **b** Picture of the etched CR39 plate. Protons with an energy between 0.8 MeV and 1.2 MeV were detected. The lower edge results from the lower aluminium foil blocking the proton beam

limited by the precision of the measured distances, especially as the magnification in this experiment was about 34-fold.

With the measured values the normalized transverse emittance was calculated using Eq. 7.8:

$$\varepsilon_{nx} \approx 5 \cdot 10^{-3}\,\pi\,\text{mm mrad}. \tag{7.8}$$

The result is similar to estimations applied in reference [2, 4]. In comparison to conventional proton accelerators with an usual emittance in the order of $1\,\pi\,$mm mrad [6] laser generated proton beams have tenfold up to hundredfold lower emittance. Therefore the proton beams are well suited for imaging experiments as presented in Part III.

References

1. S. Humphries, *Charged Particle Beams*. (Wiley, New York, 1990)
2. T.E. Cowan, J. Fuchs, H. Ruhl, A. Kemp, P. Audebert, M. Roth, R. Stephens, I. Barton, A. Blazevic, E. Brambrink, J. Cobble, J. Fernandez, J.C. Gauthier, M. Geissel, M. Hegelich, J. Kaae, S. Karsch, G.P. Le Sage, S. Letzring, M. Manclossi, S. Meyroneinc, A. Newkirk, H. Pepin, N. Renard-LeGalloudec, Ultralow emittance, multi-MeV proton beams from a laser virtual-cathode plasma accelerator. Phys. Rev. Lett. **92**(20), 204801 (2004)
3. T.E. Cowan, J. Fuchs, H. Ruhl, Y. Sentoku, A. Kemp, P. Audebert, M. Roth, R. Stephens, I. Barton, A. Blazevic, E. Brambrink, J. Cobble, J.C. Fernandez, J.C. Gauthier, M. Geissel, M. Hegelich, J. Kaae, S. Karsch, G.P. Le Sage, S. Letzring, M. Manclossi, S. Meyroneinc, A. Newkirk, H. Pepin, N. Renard-LeGalloudec, Ultra-low emittance, high current proton beams produced with a laser-virtual cathode sheath accelerator. Nucl. Instrum. Methods Phys. Res. A **544**(1–2), 277–284 (2005)
4. M. Borghesi, A.J. Mackinnon, D.H. Campbell, D.G. Hicks, S. Kar, P.K. Patel, D. Price, L. Romagnani, A. Schiavi, O. Willi, Multi-MeV proton source investigations in ultraintense laser-foil interactions. Phys. Rev. Lett. **92**(5), 055003 (2004)
5. S. Gaillard, J. Fuchs, N. Renard-Le Galloudec, T.E. Cowan, Study of saturation of CR39 nuclear track detectors at high ion fluence and of associated artifact patterns. Rev. Sci. Instrum. **78**(1), 13 (2007)
6. E. Metral, G. Franchetti, M. Giovannozzi, I. Hofmann, M. Martini, R. Steerenberg, Observation of octupole driven resonance phenomena with space charge at the CERN Proton Synchrotron. Nucl. Instrum. Methods Phys. Res. A **561**(2), 257–265 (2006)

Chapter 8
Virtual Source Dynamics

As discussed in Sect. 7.1 the model of the virtual source describes the spatial emission of the proton beam starting from a point source in front of the irradiated target. The temporal characteristics of the proton emission have been discussed theoretically and are supported by experimentally gained arguments in several publications [1–5]. The acceleration of the protons takes place on a picosecond timescale. Additionally, it was shown that the proton beam has an intrinsic velocity chirp [2, 3]. This means that energetic protons are accelerated first followed by the slower ones. For pump–probe experiments (e.g. proton imaging) the short acceleration time is more relevant since after a few millimeter the proton bunch is temporally stretched due to the broad energy distribution (cf. Sect. 9.3). Nevertheless, from the change of the emission characteristics for different proton energies, the temporal evolution of the acceleration field can be estimated even if no quantitative time scale can be given. In a previous experiment at the Max-Born-Institute [3] a fluctuation of the proton beam pointing was observed with a magnifying Thomson spectrometer. This phenomenon was also observed in the proton "streak deflectometry" experiments described in Chap. 12. Thus, a further experiment was applied, combining the advantages of both experiments—a continues energy dependent detection of several well separated parts of the proton beam.

To describe the experimental results the model of the virtual source (Sect. 7.1) is used and even further developed. Thus, information about the virtual source dynamics are obtained from these experiments. Based on this information the evolution of the acceleration field (sheath) can be estimated qualitatively in space and time.

8.1 Energy Dependent Measurement of Pinhole Projections

The purpose of the experiment was the measurement of the energy dependent evolution of the beam characteristics. Therefore the projection of a multi-pinhole

T. Sokollik, *Investigations of Field Dynamics in Laser Plasmas with Proton Imaging,* Springer Theses, DOI: 10.1007/978-3-642-15040-1_8,

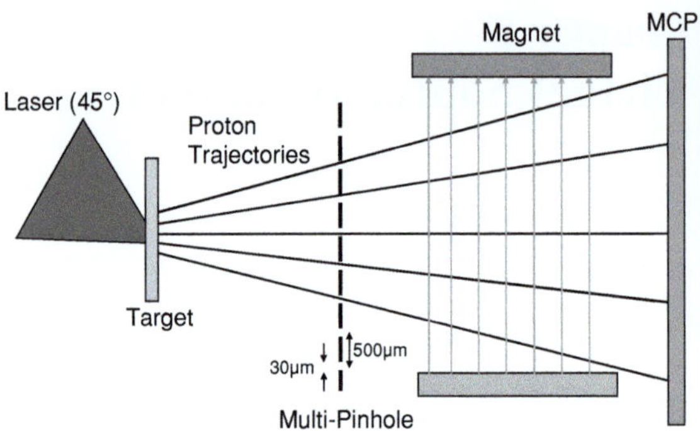

Fig. 8.1 Experimental setup. A multi-pinhole is inserted into the beam consisting of 25 pinholes in one row. The pinhole projection is detected energy sensitive due to the use of a magnet

(30 μm diameter, 500 μm spacing) was detected after the dispersive deflection by a magnet. The setup is shown in Fig. 8.1.

The resulting proton beam traces (Figs. 8.2a and 8.3a) were recorded with a MCP detection system (Sect. 9.2). Both pictures are single shot measurements with only slightly changes in the geometrical setup. The traces are elongated perpendicular to the direction of the pinhole separation due to the magnetic deflection which dependents on the proton velocity. The recorded traces fluctuate from shot to shot but the main characteristic, namely the shortening of the distance of the traces at lower proton energies, is observed in all measurements. Additionally a broadening of the distances or small scale fluctuations (both seen in Fig. 8.3a) appear in some shots. In consistence with the observations in Chap. 12 the proton beam is nearly constant at higher energies, only below ~ 0.8 MeV fluctuations appear.

The zero points represent the projection of the pinholes without deflection. They are caused by X-rays and neutral particles. Since also these signals are

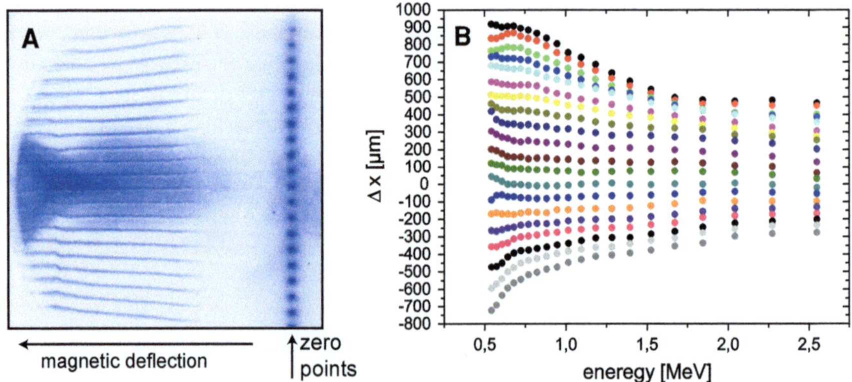

Fig. 8.2 **a** Recorded pinhole projections deflected by the magnetic field. The distance between the zero points and the proton traces define the proton energy. **b** From a readout of the traces in **a** the proton trajectories were calculated up to the target plane. The starting position Δx is plotted for different proton energies

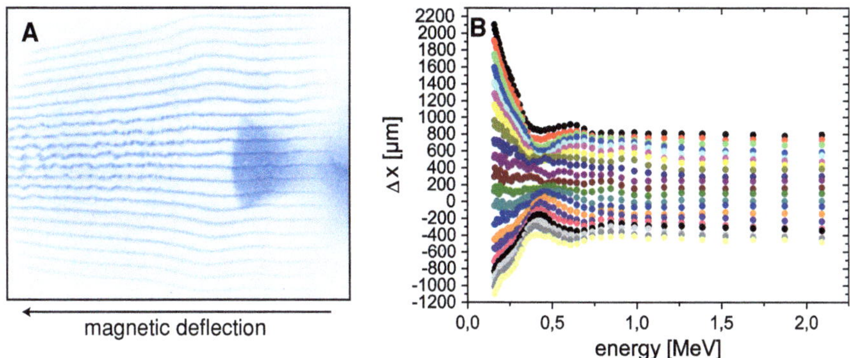

Fig. 8.3 **a** Recorded pinhole projections deflected due to the magnetic field. The zero points are at the edge of the MCP. Thus, the proton energies could still be determined. **b** The starting positions show fluctuations about 0.5 MeV, with a fast increase of the spacings at lower energies

emitted from a extended source they can be described by a virtual point source in front of the target. This fact is neglected in the following discussion since the real source has only a small extension compared to the proton source.

From the distance source to mesh (d), the distance mesh to detector (l) and the difference of the lateral position relative to the zero points (ΔX) the starting position (Δx) at the target plane can be calculated (cf. Fig. 8.4):

$$\tan \alpha = \frac{N}{d} = \frac{X}{l} \tag{8.1}$$

$$\tan \beta = \frac{X - \Delta X}{l} = \frac{N - \Delta x}{d} \tag{8.2}$$

$$\frac{N/d \cdot l - \Delta X}{l} = \frac{N - \Delta x}{d} \tag{8.3}$$

$$\Delta x = \frac{\Delta X \cdot d}{l}. \tag{8.4}$$

The calculated starting positions are plotted in Figs. 8.2b and 8.3b for both measurements. Note: The starting position Δx is determined by extrapolating the proton trajectories which are assumed as straight lines (no acting forces on the proton beam). Due to the acceleration and the high charge density of the beam at

Fig. 8.4 Schematical drawing to visualize the used variables for the calculation of the starting position Δx

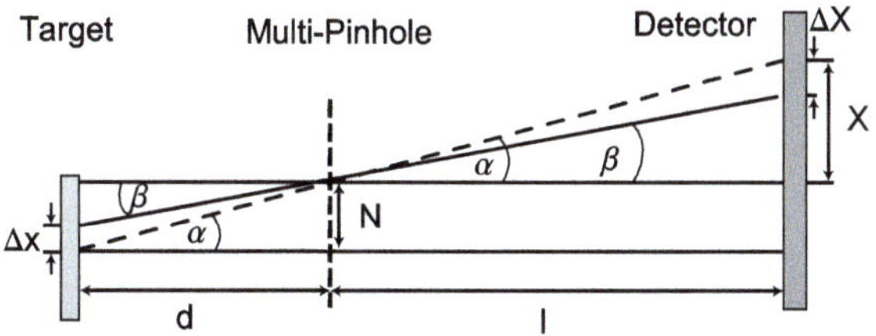

the beginning—close to the target surface—the real proton trajectories are not straight lines. This has to be taken into account if any estimation about the acceleration field and the source size will be made.

Although only a small part of the beam was detected ($\pm$ 3°) the estimated starting positions cover a relatively large area of several millimeters, especially for low proton energies. This behavior was not considered in detail in former discussions—quite recently an experiment [6] indicated such large source sizes. This confirms the observation in the imaging experiments of an electric field with an extension of several millimeters (cf. Chap. 10).

8.2 Shape of the Proton Beam

Assuming that protons with the same energy are spread continuously over a shell, the shape of this shell is defined by the proton trajectories. The trajectories are normal to the shell surface as indicated in Fig. 8.5. Up to now there is no real experimental evidence of the shape of these proton layers. In reference [7] an inverse parabolic shape is assumed in difference to other publications [8–10] which suppose a gaussian shape. In the central part of the beam both functions are nearly similar. At larger distance from the center the trajectories would cross each other in case of a gaussian shape. This would result in a ring like structure of the proton signal on the detector. Ring structures were observed in different experiments (e.g. [11]) but without the possibility to reconstruct the proton shape. Thus, it is unclear if the ring structure is caused by the gaussian shape or e.g. the saturation of the detector [12] or by magnetic fields (see Sect. 4.3).

The measurements of the previous section deliver the proton trajectories (defined by Δx and α) for the free expanding proton beam. Thus, the shape of the proton shell can be reconstructed for different proton energies. In the experiments

Fig. 8.5 The shape of the proton layer of each energy can be reconstructed from the starting positions and the angle (α) of the proton trajectories. The proton trajectories are perpendicular to the layer surface. The shape of the layer can be described by an inverse parabola or a gaussian function (*dotted line*)

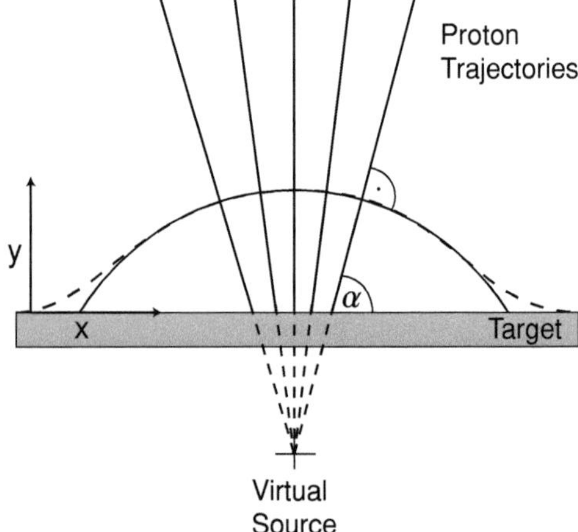

the central part of the beam ($\pm 3°$) was investigated. Thus, only the shape of proton layers in this area can be estimated. In one series, where the wings of the beam were recorded, a crossing of the pinhole projections at low energies has been observed. This is an indication of a gaussian shape. Therefore a more detailed investigation of the outer parts of the proton source is necessary. In all other measurements the central part of the beam was analyzed. In Fig. 8.6 the beam shape reconstructed from the measurement of Fig. 8.2 is shown for two different proton energies.

Therefore the angle $\tan \alpha(x)$, representing the first derivation of the "shape function", was integrated (green line, Fig. 8.6). In a second approach a parabolic function ($f(x,E_p) = 1/2\ m\ x^2$) was assumed, with the first derivation:

$$\frac{d}{dx}f(x, E_p) = \tan \alpha(x, E_p) = m(E_p) \cdot x. \tag{8.5}$$

$m(E_p)$ was determined by a linear fit of $\tan \alpha(x)$ where E_p is the proton energy. Thus, the parabolic shape of the proton layer could be reconstructed—represented in Fig. 8.2 by a black line. The red dots are the calculated positions at the target of each recorded trajectory. Thus, only their position at the abscissa (x) carries information. The value of $f(x)$ is the relative position perpendicular to the target surface. The whole beam has to be recorded to determine the absolute value. Since there are only small variations between both curves (black and green) the shape of the recorded central part of the beam can be estimated as an inverse parabola. The information which can be gained from the measurement is the continuous evolution of the proton layer curvature for different proton energies.

In Fig. 8.7a, b the curvature m of the parabolic function $f(x, E_p)$ is plotted for the recorded traces from Figs. 8.2 and 8.3, respectively. Both figures show a decreasing of the curvature at lower energies. This is astonishing since several publications report a increase of the divergence at lower proton energies [13–15]. Taking a closer look, the decreasing curvature is not in contradiction with an

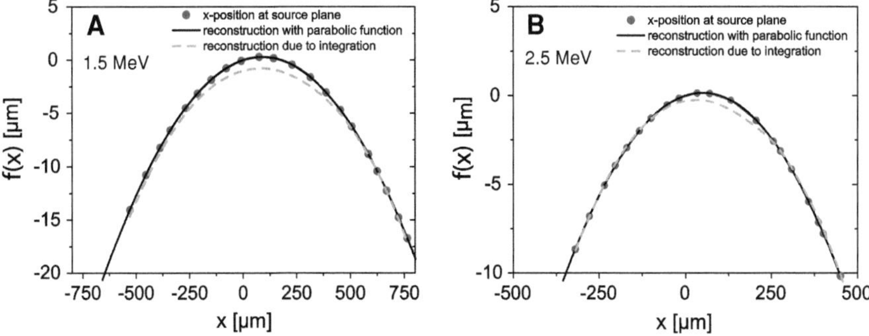

Fig. 8.6 Reconstructed shape of the proton layer for: **a** 1.5 MeV proton energy and **b** for 2.5 MeV. The *dots* indicate the starting position at the target surface. The *solid line* is an inverse parabola calculated from a linear fit of the measured angles of the trajectories. The *dashed line* is the integration of the angles dependent on the x-coordinate

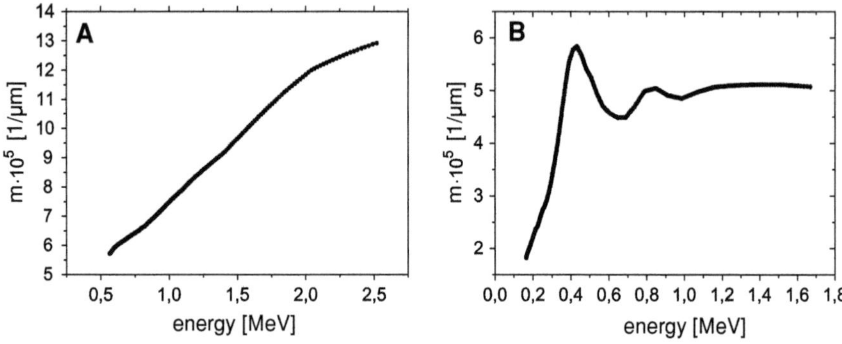

Fig. 8.7 The curvature *m* of the parabola calculated from the measurements **a**—Fig. 8.2 and **b**—Fig. 8.3

increasing divergence. The divergence $\mathrm{div}(E_p)$ is defined by the emission angle of the whole proton beam [13]:

$$\mathrm{div}(E_p) = \alpha(x_0) = \arctan\left(m(E_p) \cdot x_0\right), \tag{8.6}$$

where x_0 is the radius of the proton source and $\alpha(x_0)$ the half angle of emission. From that one can assume that the source size must grow faster than the decrease of the curvature to get an increase of the divergence at lower proton energies.

The shape of the proton layer is strongly correlated to the shape of the electron sheath (cf. Sect. 4.2). Assuming an electrostatic acceleration a prediction of the electron sheath can be made [8, 13]. Since the electric field which is responsible for the shape of the proton layer is proportional to the gradient of the electron density, the shape of the electron sheath can be described by an inverse parabolic function. These results are consistent with the findings in reference [13].

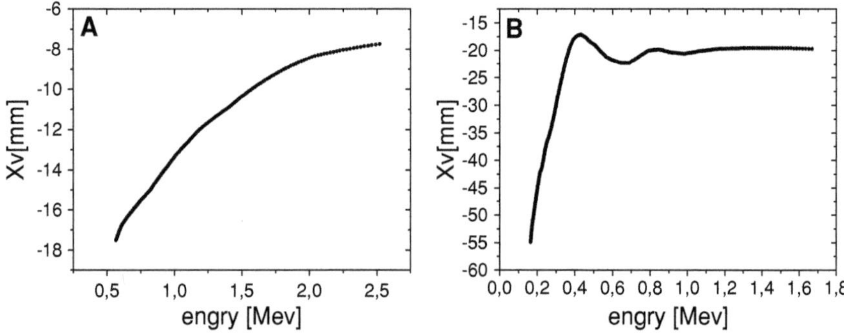

Fig. 8.8 Energy dependence of the virtual source distance x_v calculated from the measurements: **a**—Fig. 8.2 and **b**—Fig. 8.3

8.3 Energy Dependence of the Virtual Source

The energy dependence of the proton layer curvature can be expressed by the dependence of the distance of the virtual source. The crossing of the trajectories in front of the target defines its position x_v (cf. Eq. 8.5). The distance of the virtual source, determined by $x_v = 1/m(E_p)$, grows for lower proton energies (see Fig. 8.8).

The fluctuations of both measurements in Fig. 8.8 represent the behavior of all done experiments. The reasons for that are highly speculative. But with a high probability they are an imprint of the laser pulse fluctuations, for example the variation of pulse contrast and pulse energy. Thus, different absorption mechanisms can take place which influence the electron population and, hence, the proton shape. Also the curvature of the mounted target foil delivers an intrinsic curvature of the proton layers [7].

References

1. A.J. Kemp, H. Ruhl, Multispecies ion acceleration off laser-irradiated water droplets. Phys. Plasmas, **12**(3), 033105–033110 (2005)
2. T.E. Cowan, J. Fuchs, H. Ruhl, A. Kemp, P. Audebert, M. Roth, R. Stephens, I. Barton, A. Blazevic, E. Brambrink, J. Cobble, J. Fernandez, J.C. Gauthier, M. Geissel, M. Hegelich, J. Kaae, S. Karsch, G.P. Le Sage, S. Letzring, M. Manclossi, S. Meyroneinc, A. Newkirk, H. Pepin, N. Renard-LeGalloudec, Ultralow emittance, multi-MeV proton beams from a laser virtual-cathode plasma accelerator. Phys. Rev. Lett. **92**(20), 204801 (2004)
3. J. Schreiber, S. Ter Avetisyan, E. Risse, M.P. Kalachnikov, P.V. Nickles, W. Sandner, U. Schramm, D. Habs, J. Witte, M. Schnurer, Pointing of laser-accelerated proton beams. Phys. Plasmas **13**(3), 033111 (2006)
4. P.K. Patel, A.J. Mackinnon, M.H. Key, T.E. Cowan, M.E. Foord, M. Allen, D.F. Price, H. Ruhl, P.T. Springer, R. Stephens, Isochoric heating of solid-density matter with an ultrafast proton beam. Phys. Rev. Lett. **91**(12), 125004 (2003)
5. M. Borghesi, S. Kar, L. Romagnani, T. Toncian, P. Antici, P. Audebert, E. Brambrink, F. Ceccherini, C.A. Cecchetti, J. Fuchs, M. Galimberti, L.A. Gizzi, T. Grismayer, T. Lyseikina, R. Jung, A. Macchi, P. Mora, J. Osterholtz, A. Schiavi, O. Willi, Impulsive electric fields driven by high-intensity laser matter interactions. Laser Part. Beams **25**(1), 161–167 (2007)
6. P. McKenna, D.C. Carroll, R.J. Clarke, R.G. Evans, K.W.D. Ledingham, F. Lindau, O. Lundh, T. McCanny, D. Neely, A.P.L. Robinson, L. Robson, P.T. Simpson, C.G. Wahlstrom, M. Zepf, Lateral electron transport in high-intensity laser-irradiated foils diagnosed by ion emission. Phys. Rev. Lett. **98**(14), 145001 (2007)
7. E. Brambrink, J. Schreiber, T. Schlegel, P. Audebert, J. Cobble, J. Fuchs, M. Hegelich, M. Roth, Transverse characteristics of short-pulse laser-produced ion beams: a study of the acceleration dynamics. Phys. Rev. Lett. **96**(15), 154801 (2006)
8. J. Fuchs, T.E. Cowan, P. Audebert, H. Ruhl, L. Gremillet, A. Kemp, M. Allen, A. Blazevic, J.C. Gauthier, M. Geissel, M. Hegelich, S. Karsch, P. Parks, M. Roth, Y. Sentoku, R. Stephens, E.M. Campbell, Spatial uniformity of laser-accelerated ultrahigh-current MeV electron propagation in metals and insulators. Phys. Rev. Lett. **91**(25), 4 (2003)

9. L. Romagnani, J. Fuchs, M. Borghesi, P. Antici, P. Audebert, F. Ceccherini, T. Cowan, T. Grismayer, S. Kar, A. Macchi, P. Mora, G. Pretzler, A. Schiavi, T. Toncian, O. Willi, Dynamics of electric fields driving the laser acceleration of multi-MeV protons. Phys. Rev. Lett. **95**(19), 195001 (2005)

10. T.E. Cowan, J. Fuchs, H. Ruhl, Y. Sentoku, A. Kemp, P. Audebert, M. Roth, R. Stephens, I. Barton, A. Blazevic, E. Brambrink, J. Cobble, J.C. Fernandez, J.C. Gauthier, M. Geissel, M. Hegelich, J. Kaae, S. Karsch, G.P. Le Sage, S. Letzring, M. Manclossi, S. Meyroneinc, A. Newkirk, H. Pepin, N. Renard-LeGalloudec, Ultra-low emittance, high current proton beams produced with a laser-virtual cathode sheath accelerator. Nucl. Instrum. Methods Phys. Res. A Accelerators Spectrometers Detectors Assoc. Equip. **544**(1–2), 277–284 (2005)

11. E.L. Clark, K. Krushelnick, J.R. Davies, M. Zepf, M. Tatarakis, F.N. Beg, A. Machacek, P.A. Norreys, M.I.K. Santala, I. Watts, A.E. Dangor, Measurements of energetic proton transport through magnetized plasma from intense laser interactions with solids. Phys. Rev. Lett. **84**(4), 670 (2000)

12. S. Gaillard, J. Fuchs, N. Renard-Le Galloudec, T.E. Cowan, Study of saturation of CR39 nuclear track detectors at high ion fluence and of associated artifact patterns. Rev. Sci. Instrum. **78**(1), 13 (2007)

13. E. Brambrink, M. Roth, A. Blazevic, T. Schlegel, Modeling of the electrostatic sheath shape on the rear target surface in short-pulse laser-driven proton acceleration. Laser Part. Beams **24**(1), 163–168 (2006)

14. F. Lindau, O. Lundh, A. Persson, P. McKenna, K. Osvay, D. Batani, C.G. Wahlstrom, Laser-accelerated protons with energy-dependent beam direction. Phys. Rev. Lett. **95**(17), 4 (2005)

15. R.A. Snavely, M.H. Key, S.P. Hatchett, T.E. Cowan, M. Roth, T.W. Phillips, M.A. Stoyer, E.A. Henry, T.C. Sangster, M.S. Singh, S.C. Wilks, A. MacKinnon, A. Offenberger, D.M. Pennington, K. Yasuike, A.B. Langdon, B.F. Lasinski, J. Johnson, M.D. Perry, E.M. Campbell, Intense high-energy proton beams from Petawatt-laser irradiation of solids. Phys. Rev. Lett. **85**(14), 2945 (2000)

Part III
Proton Imaging

Chapter 9
Principle of Proton Imaging

As shown in Chap. 4 laser induced proton beams exhibit attributes which are well suitable for imaging purposes. The proton beams have a low longitudinal and transverse emittance. Due to the charge of the protons they are sensitive to electric and magnetic fields. Additionally, protons are scattered and stopped by interaction with matter (cf. Fig. 9.1).

Therefore proton imaging can be applied for different objectives. It can be used to visualize density gradients, for example in shocked material [2, 3] or to detect electric and magnetic fields generated by laser matter interaction (proton deflectometry) [4, 5]. With proton imaging the ion acceleration process itself [6, 7] was investigated recently. The electrostatic acceleration field and the field of the expanding ion front as well as toroidal magnetic fields [8–10] were imaged. In the following experiments done with the laser system described in Chap. 5 will be discussed which contain novel diagnostic tools and new physical findings.

9.1 Principle Experimental Setup

In Fig. 9.2 the principle setup for a proton pump-probe experiment is shown. The laser beam CPA2 irradiates the interaction target and generates a plasma at the rear side. If the intensities exceed 10^{18} W/cm^2, energetic ($\sim$MeV) ions are accelerated perpendicular to the target surface. The ion front and the expanding plasma contain electric and magnetic fields which can be probed by the proton beam generated by the CPA1 laser beam. Therefore CPA1 is focused on a thin foil with intensities $\geq 10^{19}$ W/cm^2. The resulting proton beam has a continues energy distribution with a maximum energy of several MeV, as shown in Chap. 6.

Due to the divergence of the proton beam a magnified picture of the probed object can be obtained, determined by $M = L/d$, where L is the distance between the proton source and detector, and d the distance between the source and the

T. Sokollik, *Investigations of Field Dynamics in Laser Plasmas with Proton Imaging,* Springer Theses, DOI: 10.1007/978-3-642-15040-1_9,
© Springer-Verlag Berlin Heidelberg 2011

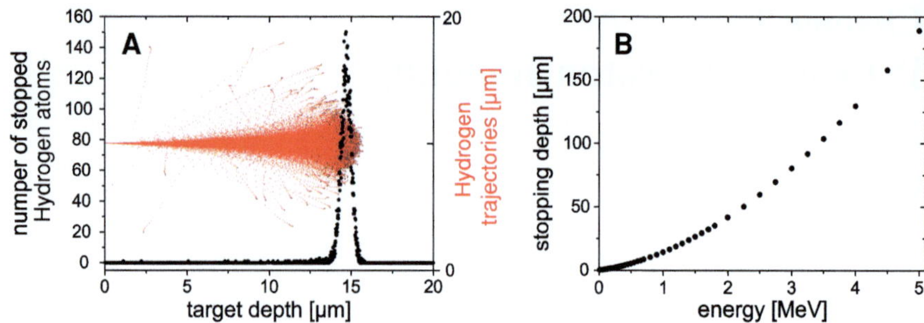

Fig. 9.1   a Trajectories and number of stopped Hydrogen atoms (1 MeV) in a 20 μm aluminum layer calculated with SRIM [1]. **b** Stopping depth in aluminum dependent on the Hydrogen energy

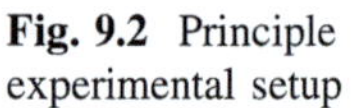

Fig. 9.2 Principle experimental setup

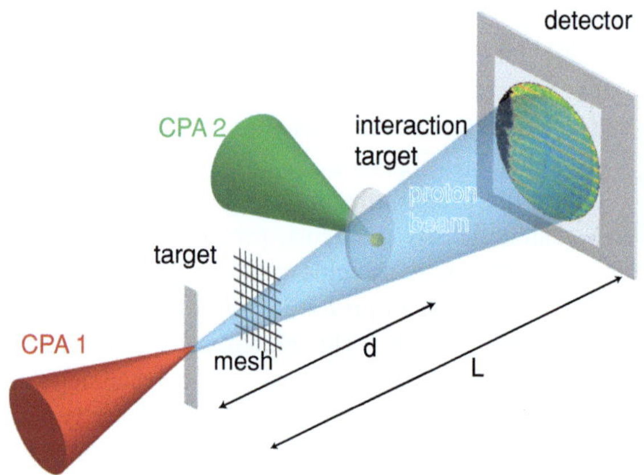

interaction target (Fig. 9.2). Usually film stacks or CR39 plates are applied to detect the protons [7], but they have several disadvantages. In single shot experiments, like the proton imaging, they have to be removed after each shot which is time consuming because the experimental chamber has to be opened and pumped down for the next shot. Even a chain of motorized stacks allows only a few shots before it is necessary to open the chamber. Another circumstance hinders the use of film stacks as they consist of several layers of radiochromic films. Protons mainly deposit energy in the Bragg-peak (see Fig. 9.1) where its position depends on the energy. According to reference [7] 50% of the signal in one layer is within an energy range of 0.5 MeV. Additionally the stacks have to be covered by an aluminum filter to prevent signals from photons.

The energy sensitive detection of the signal is highly important for temporally resolved images. The proton beam usually has a broad and continuous energy distribution—but the energetic protons are all accelerated within 1–2 ps [5, 11–14]. Starting at nearly the same time from the target, they reach the interaction target at different times. Fast (higher energetic) protons reach the object to probe earlier than slower (less energetic) protons. Hence, every exposed layer contains information of different times. The film stacks allow series of temporal snapshots, whereas the number of the exposed layers depends on the maximum energy of the proton beam. With proton energies up to 2–4 MeV, as in the present experiments at the MBI, only one or two snapshots can be recorded. For these reasons a new detecting method was applied. For the first time multi-channel plate technology was used for the specific purpose of proton imaging.

9.2 Gated Multi-Channel Plates

Multi-channel plates (MCP) (cf. Fig. 9.3) coupled to a phosphor screen and a CCD camera deliver two dimensional images with a spatial resolution of several micrometer. The detector is sensitive to energetic particles as well as to photons.

If the laser light interacts with the target electrons and x-rays are produced apart from energetic ions [15]. Since the MCP is also sensitive to these signals, the recorded images would be strongly influenced by them and not analyzable. In order to avoid this issue the MCP has to be switched on after photons and energetic electrons have reached the detector which is much earlier than the arrival of the ions.

In Fig. 9.4 the different times of flight to the detector are shown. The voltage of the MCP and the phosphor screen can be switched on and off on a time scale of several nanoseconds down to several hundreds of picoseconds with special fast high-voltage switchers. In this case the detector is only sensitive for a short time ΔT_{Gate} (gating time). This allows the measurement of a pure proton signal and additionally a selection of the energy interval of the proton beam. By choosing a

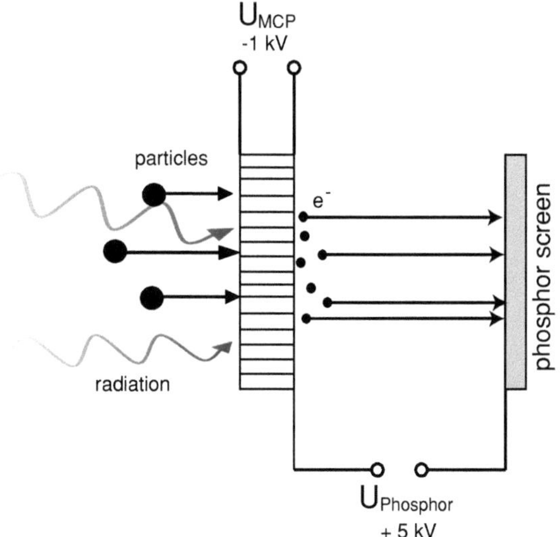

Fig. 9.3 Scheme of a MCP coupled to a phosphor screen. Multi-channel plates consist of micro-channels of several micrometer diameter. Electrons are generated when energetic particles hit the wall of the channels. Due to the voltage U_{MCP} the electrons are accelerated and generate new electrons when they hit the wall. Thus, a cascade of electrons propagates through the channel and the signal is magnified by several orders of magnitude

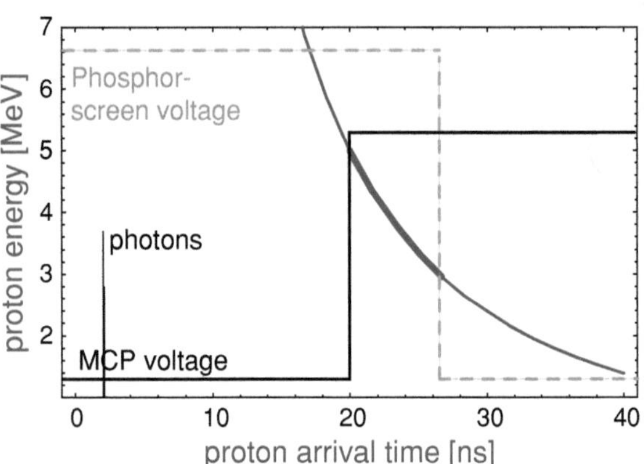

Fig. 9.4 *Solid curve* time of flight of protons from the source to the detector (L = 0.69 m). By switching the MCP voltage on (*solid step function*) the high energy cut off of the detected proton bunch can be chosen. The low energetic part of the detected bunch is defined by the switching of the Phosphor screen voltage (*dashed step function*)

specific time for the MCP gating one defines the observed time window of the interaction with the probed object. In Appendix B a detailed description of the achieved gating times and exposure times of the used gated MCP can be found.

With this technique only one snapshot per shot can be taken—but the high repetition rate allows several shots varying the time window in a few minutes. This is essentially for experiments which require a high number of measurements. Thus, the experiments in Chap. 11 were only practicable by using this method.

9.3 Time Resolution

A proton image contains of protons within an energy interval of ΔE. Due to the different energies, the proton bunch is temporally stretched (see Sect. 9.1) while propagating from the source (area of acceleration) to the object to probe. Hence the time resolution is given by the different arrival times at the object. Until reaching the detector the proton bunch is stretched further. Thus, the time resolution Δt_{Gate} depends in first approximation on the distance d to the object and the distance L to the detector (see Fig. 9.2):

$$\Delta t_{Gate} = \frac{\Delta T_{Gate}}{M}. \tag{9.1}$$

M is the magnification ($M = L/d$). Formula 9.1 represents the dominating part of the time resolution in the experiments with gated MCPs. In addition to the stretching of the proton bunch the time of flight through the object to be probed has to be taken into account. The convolution of both terms delivers the time resolution. The upper boundary (the maximum time window) is given by:

$$\Delta t \leq \Delta t_{Gate} + l\sqrt{\frac{m_p}{2E_{min}}}, \tag{9.2}$$

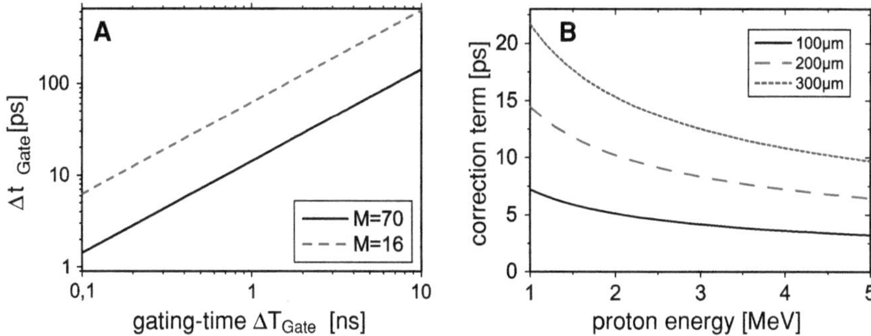

Fig. 9.5 **a** Time resolution (Δt_{Gate}) dependent on the gating time (Eq. 9.3). **b** Correction term ($l\sqrt{m_p/2E_{min}}$)—corresponds to the propagation of the protons through the object with the extension l

where l is the extension of the probed field, E_{min} the lowest energy of the detected proton bunch and m_p the proton mass. In Fig. 9.5 Δt_{Gate} and the correction term ($l\sqrt{m_p/2E_{min}}$) is plotted.

In the experiments a time resolution between 400 ps and 23 ps was achieved. The energy calibrations of the gated MCPs are discussed in Appendix B.

With this gating technique two different scenarios were investigated—the rear side plasma of thin foils (Chap. 10) and irradiated micro water-droplets (Chap. 11). In Chap. 12 an advanced method ("streak deflectometry") will be discussed which allows a continuous temporal record of the probed field with a high temporal resolution for the first time.

References

1. J.F. Ziegler, J.P. Biersack, U. Littmark, *Stopping and Range of Ions in Solids*. (Pergamon Press, New York, 1996)
2. M. Koenig, E. Henry, G. Huser, A. Benuzzi-Mounaix, B. Faral, E. Martinolli, S. Lepape, T. Vinci, D. Batani, M. Tomasini, B. Telaro, P. Loubeyre, T. Hall, P. Celliers, G. Collins, L. DaSilva, R. Cauble, D. Hicks, D. Bradley, A. MacKinnon, P. Patel, J. Eggert, J. Pasley, O. Willi, D. Neely, M. Notley, C. Danson, M. Borghesi, L. Romagnani, T. Boehly, K. Lee, High pressures generated by laser driven shocks: applications to planetary physics. Nucl. Fusion **44**(12), S208–S214 (2004)
3. A.J. Mackinnon, P.K. Patel, R.P. Town, M.J. Edwards, T. Phillips, S.C. Lerner, D.W. Price, D. Hicks, M.H. Key, S. Hatchett, S.C. Wilks, M. Borghesi, L. Romagnani, S. Kar, T. Toncian, G. Pretzler, O. Willi, M. Koenig, E. Martinolli, S. Lepape, A. Benuzzi-Mounaix, P. Audebert, J.C. Gauthier, J. King, R. Snavely, R.R. Freeman, T. Boehlly, Proton radiography as an electromagnetic field and density perturbation diagnostic (invited). Rev. Sci. Instrum. **75**(10), 3531–3536 (2004)

4. M. Borghesi, D.H. Campbell, A. Schiavi, M.G. Haines, O. Willi, A.J. Mackinnon, P. Patel, L.A. Gizzi, M. Galimberti, R.J. Clarke, F. Pegoraro, H. Ruhl, S. Bulanov, Electric field detection in laser-plasma interaction experiments via the proton imaging technique. Phys. Plasmas 9(5), 2214–2220 (2002)

5. M. Borghesi, S. Kar, L. Romagnani, T. Toncian, P. Antici, P. Audebert, E. Brambrink, F. Ceccherini, C.A. Cecchetti, J. Fuchs, M. Galimberti, L.A. Gizzi, T. Grismayer, T. Lyseikina, R. Jung, A. Macchi, P. Mora, J. Osterholtz, A. Schiavi, O. Willi, Impulsive electric fields driven by high-intensity laser matter interactions. Laser Part. Beams 25(1), 161–167 (2007)

6. L. Romagnani, J. Fuchs, M. Borghesi, P. Antici, P. Audebert, F. Ceccherini, T. Cowan, T. Grismayer, S. Kar, A. Macchi, P. Mora, G. Pretzler, A. Schiavi, T. Toncian, O. Willi, Dynamics of electric fields driving the laser acceleration of multi-MeV protons. Phys. Rev. Lett. 95(19), 195001 (2005)

7. M. Borghesi, A. Schiavi, D.H. Campbell, M.G. Haines, O. Willi, A.J. Mackinnon, P. Patel, M. Galimberti, L.A. Gizzi, Proton imaging detection of transient electromagnetic fields in lasor-plasma interactions (invited). Rev. Sci. Instrum. 74(3), 1688–1693 (2003)

8. C.K. Li, F.H. Seguin, J.A. Frenje, J.R. Rygg, R.D. Petrasso, R.P.J. Town, P.A. Amendt, S.P. Hatchett, O.L. Landen, A.J. Mackinnon, P.K. Patel, V.A. Smalyuk, T.C. Sangster, J.P. Knauer, Measuring E and B fields in laser-produced plasmas with monoenergetic proton radiography. Phys. Rev. Lett. 97(13), 135003 (2006)

9. C.K. Li, F.H. Seguin, J.A. Frenje, J.R. Rygg, R.D. Petrasso, R.P.J. Town, P.A. Amendt, S.P. Hatchett, O.L. Landen, A.J. Mackinnon, P.K. Patel, M. Tabak, J.P. Knauer, T.C. Sangster, V.A. Smalyuk, Observation of the decay dynamics and instabilities of megagauss field structures in laser-produced plasmas. Phys. Rev. Lett. 99(1), 015001 (2007)

10. S. Le Pape, D. Hey, P. Patel, A. Mackinnon, R. Klein, Proton radiography of megagauss electromagnetic fields generated by the irradiation of a solid target by an ultraintense laser pulse. Astrophys. Space Sci. 307(1–3), 341–345 (2007)

11. A.J. Kemp, H. Ruhl, Multispecies ion acceleration off laserirradiated water droplets. Phys. Plasmas 12(3), 033105–10 (2005)

12. T.E. Cowan, J. Fuchs, H. Ruhl, A. Kemp, P. Audebert, M. Roth, R. Stephens, I. Barton, A. Blazevic, E. Brambrink, J. Cobble, J. Fernandez, J.C. Gauthier, M. Geissel, M. Hegelich, J. Kaae, S. Karsch, G.P. Le Sage, S. Letzring, M. Manclossi, S. Meyroneinc, A. Newkirk, H. Pepin, N. Renard-LeGalloudec, Ultralow emittance, multi-MeV proton beams from a laser virtual-cathode plasma accelerator. Phys. Rev. Lett. 92(20), 204801 (2004)

13. J. Schreiber, S. Ter Avetisyan, E. Risse, M.P. Kalachnikov, P.V. Nickles, W. Sandner, U. Schramm, D. Habs, J. Witte, M. Schnurer, Pointing of laser-accelerated proton beams. Phys. Plasmas 13(3), 033111 (2006)

14. P.K. Patel, A.J. Mackinnon, M.H. Key, T.E. Cowan, M.E. Foord, M. Allen, D.F. Price, H. Ruhl, P.T. Springer, R. Stephens, Isochoric heating of solid-density matter with an ultrafast proton beam. Phys. Rev. Lett. 91(12), 125004 (2003)

15. S.P. Hatchett, C.G. Brown, T.E. Cowan, E.A. Henry, J.S. Johnson, M.H. Key, J.A. Koch, A.B. Langdon, B.F. Lasinski, R.W. Lee, A.J. Mackinnon, D.M. Pennington, M.D. Perry, T.W. Phillips, M. Roth, T.C. Sangster, M.S. Singh, R.A. Snavely, M.A. Stoyer, S.C. Wilks, K. Yasuike, Electron, photon, and ion beams from the relativistic interaction of Petawatt laser pulses with solid targets. Phys. Plasmas 7(5), 2076–2082 (2000)

Chapter 10
Imaging Plasmas of Irradiated Foils

To accelerate protons up to energies of several MeV, mostly thin foils are used. They deliver a homogeneous beam with an emission angle of about 20° (half angle). The study of the acceleration process plays the key role in understanding fundamental physics and delivers possibilities to manipulate the beam characteristics for future applications. Also the expansion of the plasma is an important issue being in the focus of recent investigations [1, 2]. By probing the rear side of an irradiated foil one can gain information about the expanding plasma and the acceleration fields.

10.1 Experimental Setup

The experimental setup is shown in Fig. 10.1. To generate the proton probe beam the Ti:Sa laser was focused at a 12 μm aluminum foil reaching intensities of about 10^{19} W/cm^2. The plasma which is to be probed is created at a curved aluminum foil of 12 μm thickness and 5 mm curvature radius. Therefore the interaction target was irradiated by the Nd:glass laser with an intensity between 10^{17} and 10^{18} W/cm^2. Both lasers were synchronized electronically with a precision of several picoseconds (see Fig. 5.9). The time delay between the laser pulses was adjusted using an optical delay stage. The time of flight of the protons from the source to the interaction target was taken into account for the correct timing. A magnification of 16-fold was achieved in the proton images related to the plane of the interaction target (L = 0.604 m, d = 0.04 m, cf. Fig. 10.1). A mesh with 500 lpi (lines per inch) placed 30 mm away from the target intersects the proton beam into small beamlets. This allows a better reconstruction of the deflecting fields. The proton probe beam was detected with a gated multi-channel plate (MCP) coupled to a phosphor screen.

T. Sokollik, *Investigations of Field Dynamics in Laser Plasmas with Proton Imaging,* Springer Theses, DOI: 10.1007/978-3-642-15040-1_10,
© Springer-Verlag Berlin Heidelberg 2011

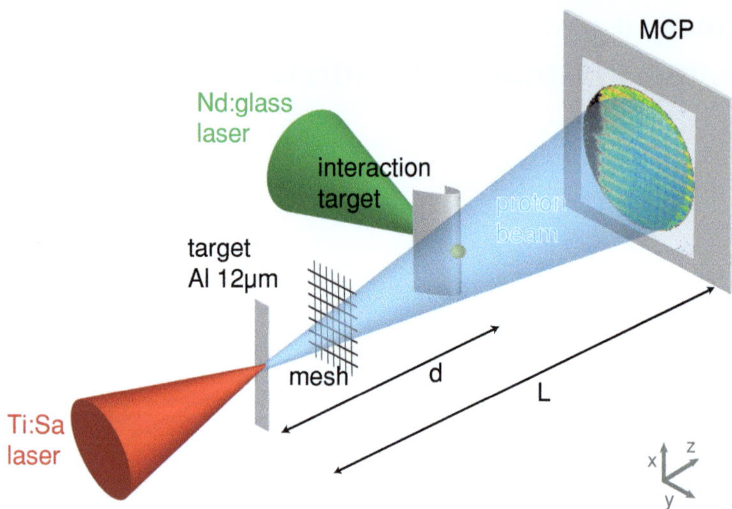

Fig. 10.1 Experimental setup

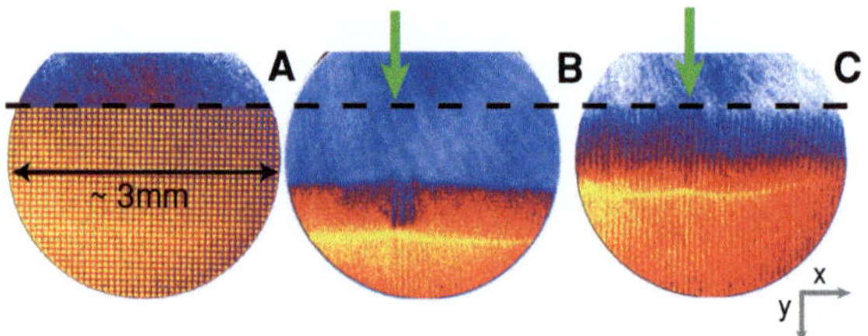

Fig. 10.2 Temporal snapshots of the target edge (*upper boundary of picture*) with a proton beam intersected by a mesh: **a** interaction target not exposed, **b, c** interaction target irradiated at 10^{18} W/cm^2, proton image produced with (1.4–2) MeV protons due to MCP gating. The *arrow* indicates CPA2 laser irradiation on the interaction target

10.2 2D-Proton Images

Proton images of the rear side of the foils are shown in Fig. 10.2, whereas Fig. 10.2a is an undisturbed picture. The upper boundary results from the edge of the interaction target which blocks the protons. In this experiment a mesh was placed inside the beam. Since there is no interaction with the Nd:glass laser pulse the mesh structure is clearly visible. If the laser irradiates the target (Fig. 10.2b, c) the proton beam is deflected away from the target surface over the full observation area ($\sim$3 mm). Additionally a distortion at the laser interaction point is visible in Fig. 10. 2b which shows a magnified mesh structure. A yellow line in Fig. 10.2b, c indicating an area of an enhanced number of protons. The pictures are time integrated over 400 ps due to the MCP gating. Circular magnetic fields which are generated perpendicular to the target surface [3–5] can be neglected in this imaging scenario [6].

To explain the observed structures in the images the processes caused by the interaction with the intense laser pulse have to be taken into account (see Sect. 3.3 and Chap. 4). Electrons are accelerated to relativistic velocities. They propagate through the target spreading with an angle of approximately 10°–30° [7, 8] and leave the target at the rear side. Some of these hot electrons, named precursor electrons, escape leaving a positive charged target behind. Electrons which can not escape are trapped by the rising electric potential and create the electron sheath. This electron cloud spreads over the target surface while cold electrons of the target foil built up a return current [3, 9]. The resulting electric field ionizes the contamination layer at the target surface and accelerates the ions.

The initial acceleration field, with the smallest extension of the electron sheath, accelerates ions to the highest energies. When the sheath expands laterally (radially) the field strength decreases rapidity. In several proton images this initial field can be observed (e.g. Figs. 10.2b and 10.3b). A bell like structure is visible and the maximum deflection from the target surface is defined by the strength and the extension of the field. While expanding the field strength decreases. Thus, the deflection of the probe beam has a similar value as in the initial case. Hence the maximal deflection by this field is visible in the proton images in form of a line structure with the same distance to the target surface as the bell like structure. The sheath field acts only on protons of the probe beam in a short energy interval. Hence a distinct line structure is visible. This line of enhanced proton numbers is a time integrated result of the acceleration field which expands radially.

In other words—energetic ions are accelerated from the rear side with a source size much larger than the focal spot of the laser because of the spreading of the electrons. Due to the spreading of the electrons over the rear surface lower energetic ions are accelerated from an area up to a few millimeter. This observation was recently confirmed by reference [9] and by the experiments in Chap. 8.

The expanding ion front itself propagates over the observed area and blurs the mesh structure in y-direction (cf. Fig. 10.2). Between target surface and front the field of the positively charged target becomes visible in the pictures.

In Fig. 10.3 a series of snapshots is shown. The delay between the Nd:glass laser and the Ti:Sa laser was changed with an optical delay stage in the beamline of the Nd:glass laser. The gating time for the proton detection was held constant. The fields at the interaction target interact at different times with the proton bunch—hence with protons of different energies. Also the Nd:glass laser energy was varying from shot to shot. The line of enhanced proton numbers shows sometimes a folded structure. A possible explanation therefore is given in reference [9]. While the electron cloud expands the field strength is locally enhanced dependent on the time of the expansion.

For a quantitative analysis of the field strength the deflection by the sheath field was measured (distance of the line to the target surface). Assuming a effective

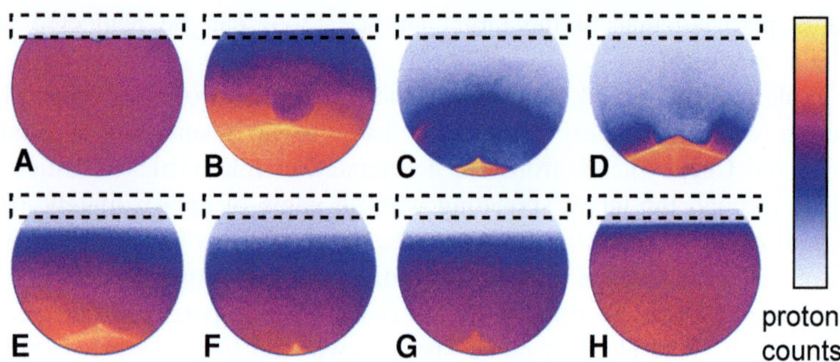

Fig. 10.3 2D-snapshots of proton beam deflection at different observation times in respect to the irradiation of the CPA2-pulse which creates the deflection field: **a** cold target, **b** (0 ± 200) ps, **c** (200 ± 200) ps, **d** (400 ± 200) ps, **e** (600 ± 200) ps, **f** (700 ± 200) ps, **g** (800 ± 200) ps, **h** (900 ± 200) ps. The integration time interval is defined by the MCP gating (cf. text), the color codes the proton number as given in the legend

electric field which is homogenously and temporally constant perpendicular to the target surface (in y-direction) the deflection (Y) depends on the energy of the protons E_p, the field strength E_{el} and the extension of the field l. The proton energy was determined roughly by the time when the proton image gets disturbed (changing the delay of the Nd:glass laser).

The acting force in y-direction is defined by:

$$E_{el} \cdot e = m_p \cdot \frac{d^2 y}{dt^2}.$$

(10.1)

Using $dt = 1/v_z\, dz$ leads to:

$$v_y = \frac{dy}{dt} = \frac{e}{m_p \cdot v_z} \int E_{el}\, dz.$$

(10.2)

With the effective field extension l the velocity is given by:

$$v_y = \frac{e \cdot E_{el} \cdot l}{m_p \cdot v_z}.$$

(10.3)

The deflection at the detector can be estimated by:

$$Y = \frac{v_y}{v_z} L,$$

(10.4)

where L is the distance between interaction target and detector and v_z the velocity of the proton in propagation direction z. Measuring the deflection at the detector for known proton energies delivers the field strength for an assumed extension l:

$$E_{el} = \frac{2 Y E_p}{e l L}.$$

(10.5)

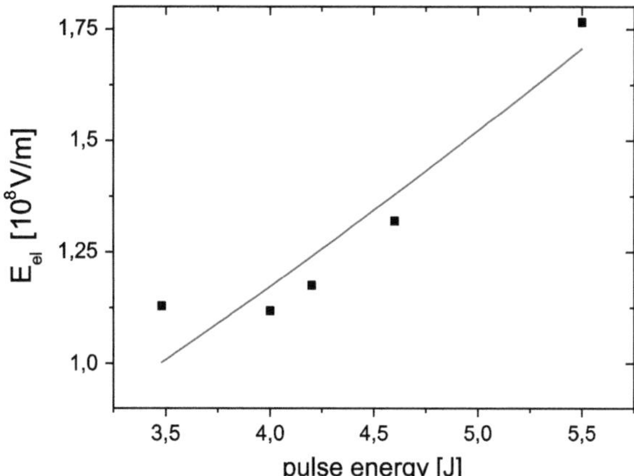

Fig. 10.4 Estimated field strength over pulse energy of CPA2 (*rectangles*). The *solid line* indicates the dependence of $E_{el} \propto \sqrt{E_{\text{laser}}}$

In Fig. 10.4 the estimated electric field which is responsible for the creation of the line structure is plotted for different pump energies. The effective extension of the field was assumed to be $l = 1$ mm.

The estimated field strength represents an average over time and space. The dependence on the square root of the pulse energy is indicated in Fig. 10.4 and fits with theoretical scaling laws [10]. The low time resolution in the experiments prevent further statements about the temporal evolution of the fields. In Chap. 12 the dynamical processes are investigated in more detail by using an novel imaging technique.

References

1. P. Mora, Plasma expansion into a vacuum. Phys. Rev. Lett. **90**(18), 185002 (2003)
2. P. Mora, Thin-foil expansion into a vacuum. Phys. Rev. E **72**(5), 5 (2005)
3. T. Nakamura, S. Kato, H. Nagatomo, K. Mima, Surface-magnetic-field and fast-electron current-layer formation by ultraintense laser irradiation. Phys. Rev. Lett. **93**(26), 4 (2004)
4. M. Zepf, E.L. Clark, F.N. Beg, R.J. Clarke, A.E. Dangor, A. Gopal, K. Krushelnick, P.A. Norreys, M. Tatarakis, U. Wagner, M.S. Wei, Proton acceleration from high-intensity laser interactions with thin foil targets. Phys. Rev. Lett. **90**(6), 064801 (2003)
5. Y. Murakami, Y. Kitagawa, Y. Sentoku, M. Mori, R. Kodama, K.A. Tanaka, K. Mima, T. Yamanaka, Observation of proton rear emission and possible gigagauss scale magnetic fields from ultra-intense laser illuminated plastic target. Phys. Plasmas **8**(9), 4138–4143 (2001)
6. C.K. Li, F.H. Seguin, J.A. Frenje, J.R. Rygg, R.D. Petrasso, R.P.J. Town, P.A. Amendt, S.P. Hatchett, O.L. Landen, A.J. Mackinnon, P.K. Patel, V.A. Smalyuk, T.C. Sangster, J.P. Knauer, Measuring E and B fields in laser-produced plasmas with monoenergetic proton radiography. Phys. Rev. Lett. **97**(13), 135003 (2006)
7. A.A. Andreev, R. Sonobe, S. Kawata, S. Miyazaki, K. Sakai, K. Miyauchi, T. Kikuchi, K. Platonov, K. Nemoto, Effect of a laser prepulse on fast ion generation in the interaction of ultra-short intense laser pulses with a limited-mass foil target. Plasma Phys. Control. Fusion **48**(11), 1605–1619 (2006)
8. F. Amiranoff, Fast electron production in ultra-short high-intensity laser-plasma interaction and its consequences. Meas. Sci. Technol. **12**(11), 1795–1800 (2001)

9. P. McKenna, D.C. Carroll, R.J. Clarke, R.G. Evans, K.W.D. Ledingham, F. Lindau, O. Lundh, T. McCanny, D. Neely, A.P.L. Robinson, L. Robson, P.T. Simpson, C.G. Wahlstrom, M. Zepf, Lateral electron transport in high-intensity laser-irradiated foils diagnosed by ion emission. Phys. Rev. Lett. **98**(14), 145001 (2007)
10. S.P. Hatchett, C.G. Brown, T.E. Cowan, E.A. Henry, J.S. Johnson, M.H. Key, J.A. Koch, A.B. Langdon, B.F. Lasinski, R.W. Lee, A.J. Mackinnon, D.M. Pennington, M.D. Perry, T.W. Phillips, M. Roth, T.C. Sangster, M.S. Singh, R.A. Snavely, M.A. Stoyer, S.C. Wilks, K. Yasuike, Electron, photon, and ion beams from the relativistic interaction of Petawatt laser pulses with solid targets. Phys. Plasmas **7**(5), 2076–2082 (2000)

Chapter 11
Mass-Limited Targets

In the last decade laser induced ion acceleration became an attractive source for energetic ion beams up to several MeV (max. 58 MeV [1]). For further applications besides the proton imaging, new parameter ranges have to be occupied. The proton energies have to be increased about one order of magnitude. Many applications would benefit from a narrow energy spread of the protons—so called "monoenergetic" protons beams. But also the number of protons has to be increased reproducibly.

Aside from increasing the laser intensity and contrast, which is a challenging and cost-intensive task, the choice of the target system has the most promising potential. Recent investigations show that the ion beam can be manipulated by the use of special targets. Concave targets focus or collimate the whole ion beam [2–5]. Manipulated target surfaces or micro-structured targets can deliver quasi-monoenergetic ion and proton beams [6, 7]. At the Max-Born-Institute it was shown for the first time that also water-droplet targets can deliver quasi-monoenergetic deuteron and proton beams [8, 9] (see Sect. 6.2).

Water droplets are a very promising target system. They can deliver monoenergetic protons without the necessity of further manipulation of the target. They can be used with a high repetition rate—only limited by the repetition rate of current laser system. And it has been shown that the laser energy conversion is more efficient than at other target systems [10]. Theoretical investigations also predict higher proton energies [11]—but this has not been observed until now. Unlike all other common targets, water-droplets are spherical, mass limited and not grounded.

One disadvantage seems to be the emission direction. Due to the spherical geometry the protons are usually emitted over the full solid angle. Dependent on the laser parameters (e.g. contrast) and the interaction point (central or non-central hit of the droplet) the emission can be slightly enhanced in different directions [12–14]. However the experiments where the monoenergetic feature was observed gave the first hint of a non-isotropic emission of the ions. This feature was only

T. Sokollik, *Investigations of Field Dynamics in Laser Plasmas*
with Proton Imaging, Springer Theses, DOI: 10.1007/978-3-642-15040-1_11,
© Springer-Verlag Berlin Heidelberg 2011

observed in laser forward direction. Since such a directed and monoenergetic ion beam which can be produced with a high repetition rate is of high interest for applications, this mechanism needs to be investigated in more detail. Therefore proton imaging was used for the first time to study this promising target system for proton acceleration.

11.1 Experimental Setup

The experiments have been carried out with the Ti:Sa laser (see Chap. 5). The beam was divided in two beams CPA1 (0.6 J) and CPA2 (0.15 J) with 40 fs and 60 fs, respectively. The longer pulse duration of CPA2 is caused by the propagation through the beam-splitter. The dispersion of the quartz glass elongates the pulse. The laser pulse was characterized by a SPIDER [15] measurement.

CPA1 was focused on a 5 μm thin Titanium foil (1×10^{19} W/cm^2) generating a proton beam which was used to probe the plasma—created by CPA2 (2×10^{18} W/cm^2) on 16 μm water droplets (see Fig. 11.1).

The water droplets were produced with a commercial pulsed nozzle from Siemens-Elema (Sweden) forming a chain of droplets with a droplet distance of ~ 45 μm (see Sect. 11.2). The proton probe beam was detected with a gated MCP-system which allows the selection of a narrow energy interval of the beam (gating time 5–10 ns). The magnification of the resulting proton image is about 70-fold related to the droplets, determined by $M = L/d$, where L is the distance of the

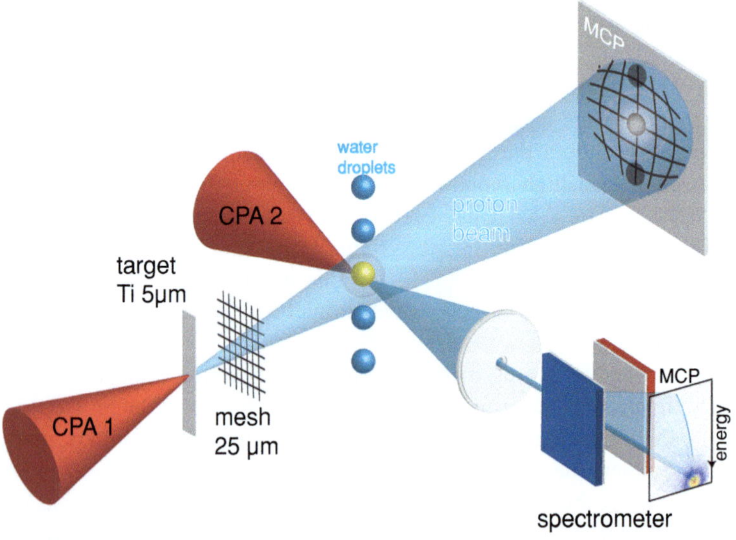

Fig. 11.1 Experimental setup. Proton imaging scheme combined with a Thomson spectrometer to measure the proton energies of the emitted protons from the irradiated droplet

proton source to the detector ($L = 1.05$ m) and d the distance of the source to the droplet ($d = 0.015$ m). In some experimental series a mesh was inserted 9 mm behind the first target. The nickel mesh consists of 1,000 lines per inch. Protons and ions emitted from the water droplet were detected simultaneously by a Thomson spectrometer [16].

11.2 Water Droplet Generation

The water droplets are generated by a small glass capillary of about 10 μm diameter. Water is pressed through the capillary generating a fine water jet (Fig. 11.2).

Due to surface tension, acting against the inertia of the fluid the jet breaks up into irregularly sized and irregularly spaced droplets (Fig. 11.3) [17]. For a regular chain of droplets the jet has to be modulated with a piezo element included in the nozzle. The high frequency modulations (0.9–1.3 MHz) cause a pressure modulation which leads to a modulation of the jet diameter [18]. Thus, the jet breaks up into a regularly chain of droplets (Fig. 11.3).

The droplet size depends on the background pressure and the modulating frequency [18, 19]. The background pressure is the pressure with which the water is pushed through the capillary. In the presented experiments the background pressure was held fixed at 30 bar. With a modulating frequency of 1.2 MHz the droplet size is approximately 16 μm with a spacing of about 45 μm. The velocity of the droplets can be estimated, knowing the distance of the droplets and the modulating frequency to be about 54 m/s.

After a few millimeter the droplets fall into a tube with an aperture of 3 mm. This tube was connected with a heated bowl to avoid the creation of ice stalagmites which would grow up to the nozzle. This separated vacuum apparatus was pumped by two scroll pumps. A cooling trap was installed additionally to reduce

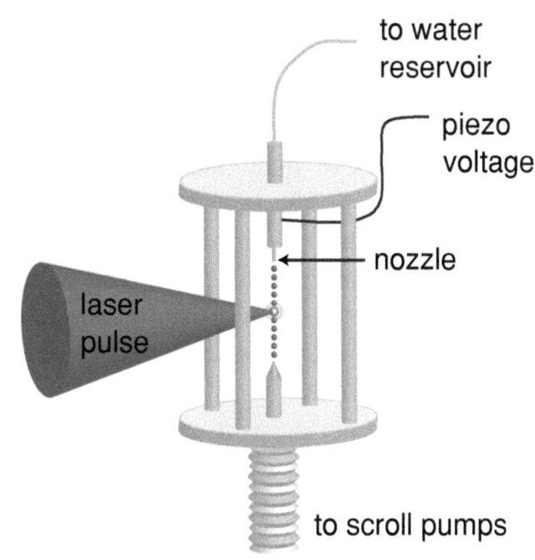

Fig. 11.2 Schematic setup of the droplet generation. The nozzle connected to a piezo element is mounted in an aluminium frame. The laser pulse interacts with the water droplet several millimeters below the capillary. The droplets are collected by a tube which is connected to scroll pumps and a cooling trap

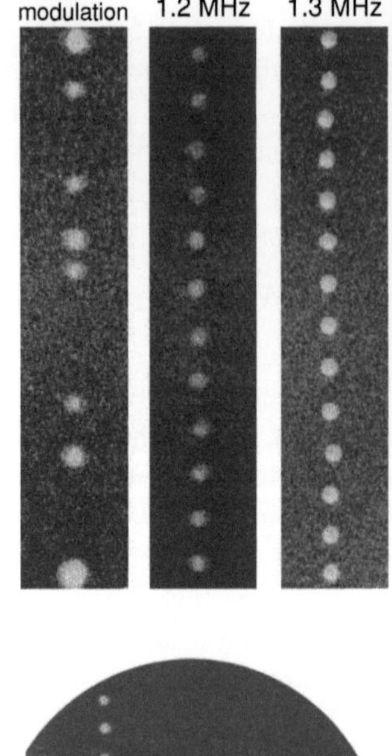

Fig. 11.3 Proton images of the droplet chain. Without modulation, with a modulation frequency of 1.2 and 1.3 MHz. The size of the droplets is about 16 μm with a spacing of 45 and 42 μm, respectively

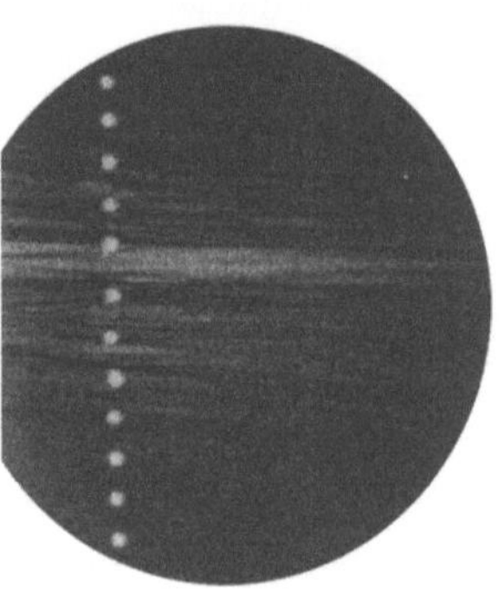

Fig. 11.4 In this proton image the laser pulse misses the droplet. The filaments are caused by the interaction of the laser pulse with the background gas (laser from the left)

the pressure down to 10^{-2} mbar. Due to the use of several turbo pumps and cooling traps the background pressure in the experimental chamber was 10^{-5} mbar. To achieve the necessary pressure of 10^{-6} mbar for the MCP, the way to the imaging MCP-detector was pumped differentially. Even if the background pressure was very low the local gas density near the droplet chain seems to be several orders of magnitude higher. In Fig. 11.4 a proton image of the droplet chain is shown. The laser (CPA2) (from the left side) misses the droplets. Nevertheless, the proton probe beam is influenced. This cannot be caused by a direct interaction with the laser pulse. The laser pulse ionizes the gas around the droplet and the droplets surface by the wings of the pulse. While propagating through the plasma the laser breaks into filaments and creates magnetic fields which deflect the proton beam [20].

The surrounding gas is caused by the evaporation of the water into vacuum and the evaporation of the droplets due to the incident laser light. The ambient pressure was estimated roughly to be 10^{-3} to 10^{-4} mbar in the vicinity of a few 100 μm of the droplets. Due to this fact the irradiated droplet can not be regarded as a fully isolated system. The surrounding gas is ionized by the laser and can contribute to the charge compensation of the droplet (see Sect. 11.3).

11.3 Proton Images of Irradiated Water Droplets

In contrast to flat foils the spherical shape of the water droplets of about 15–20 μm diameter limits the lateral electron transport [21]. Electrons are accelerated by the laser pulse—penetrate into the droplet and leave the droplet at the rear side. They escape until the resulting potential of the positive charged droplet prohibit further migration. Electrons which can not escape built up an electrostatic field (sheath) at the rear side, perpendicular to the target surface which accelerates the protons. This is the usual TNSA-mechanisms (see Chap. 4). Since the droplet is not grounded directly the sheath spreads around the droplet while decreasing. Due to the spherical symmetry protons are accelerated over the full solid angle while modifications of isotropic emission have been observed [12–14]. Furthermore, electrons produced at the front side can reach the rear side by passing around the droplets which may lead to an additional enhancement of the electrostatic field on the rear side [8].

In the experiments a radially symmetric deflection was observed caused by the positive charged droplet. In addition an asymmetric deflection of the proton beam was recorded at the rear side of the droplets. In Fig. 11.5 a symmetric deflection of the probe beam is shown. This images is time integrated over 150 ps including the beginning of the interaction of the laser with the droplet. By shortening the gating time and thus improving the time resolution of the proton images, fine structures become visible.

A distinguished feature in almost all images of Fig. 11.6 are fine filaments which are perpendicular to the target surface. A possible explanation for these structures are Weibel-like instabilities caused by counter streaming electron currents [22]. Hot electrons which cannot overcome the electrostatic barrier return into the droplet. Inside the target the cold electrons built up a counter streaming

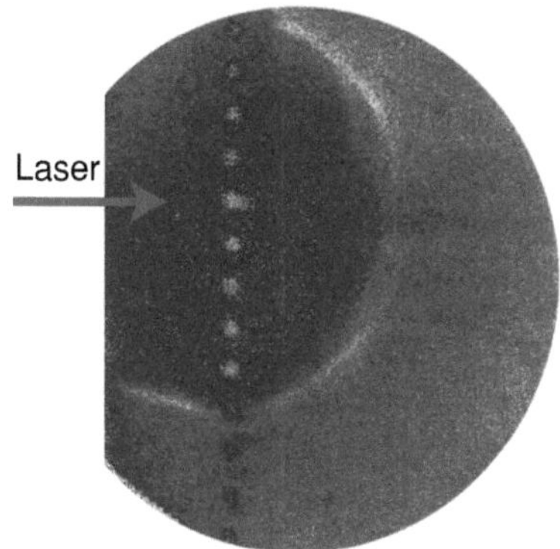

Fig. 11.5 Radial symmetrical deflection of the proton beam (time resolution ∼150 ps)

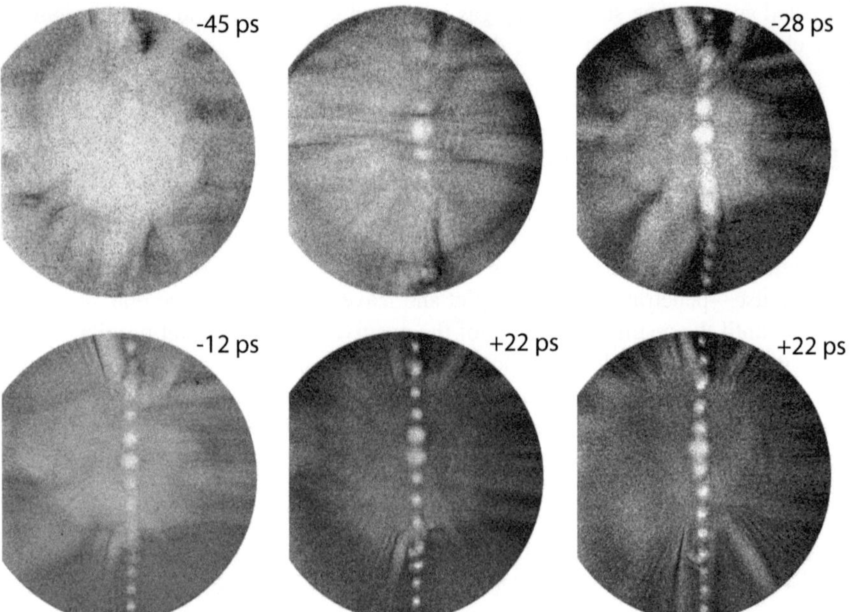

Fig. 11.6 A series of irradiated droplets (exposure time about 65 ps). Showing a spherical emission and fine filament structures caused by Weibel-like instabilities ($t = \pm 32$ ps)

current [23]. This unstable regime leads to filaments at the target surface connected to surrounding magnetic fields and a radial electric field which is balancing the electron pressure [23]. Due to the plasma expansion the imprinted structures extend over several hundred μm.

The temporal evolution of the proton deflection is shown in Fig. 11.7. Starting from an undistorted picture at a time before the laser hits the droplet, strong fields lead to a deflection of the proton probe beam at later times ($t > 0$ ps, where $t = 0$ represents the time when CPA2 hits the droplet). In this picture ($t = 0$ ps) strong fields are visible even if no mesh was inserted. The image of the droplet chain is strongly distorted. Thus, the probe beam is not stopped by matter, it is deflected by electric fields. Later on, the images of the droplets are not distorted, except of the irradiated droplet. The protons are stopped and scattered by the expanding droplet.

A mesh was inserted into the proton beam for detailed investigations. The results are shown in the upper row of Fig. 11.8. The symmetric deflection of the probe beam is clearly visible. The mesh structure is distorted similarly to the deflection caused by a Coulomb potential. Thus, the observed deflection can be associate with the charged droplet caused by the leaving precursor electrons.

The charge of the droplet can be roughly estimated analytically. Using the scaling laws given in reference [24], one can calculate the hot electron temperature T_h of the electrons produced by the CPA2 laser pulse:

$$T_h \approx m_e c^2 (\gamma - 1), \tag{11.1}$$

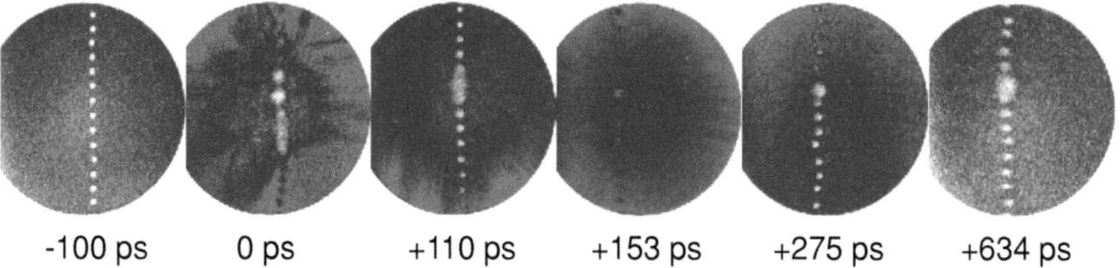

Fig. 11.7 Proton images of the droplets at different probing times within a time window of $\pm$ 32.5 ps. (laser from the left)

Fig. 11.8 In the *upper row* a mesh was inserted in the proton beam (exposure time 65 ps, $t\sim 0 \pm 32.5$ ps). In the *lower row* the mesh was removed and the exposure time was about 150 ps ($t\sim 0 \pm 75$ ps). Laser from the left. At the rear side in laser direction a broad channel-like structure is visible indicating a directed ion emission

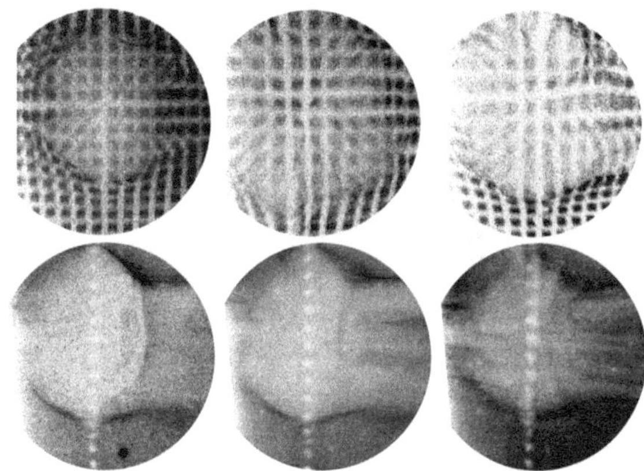

where $\gamma = (1 + 0.7I_{18}\lambda^2_{L,\mu m})^{1/2}$ with m_e—the electron rest mass, I_{18}—the laser intensity in units of 10^{18} W/cm^2 and $\lambda_{L,\mu m}$—the laser wavelength in μm. The energetic electrons propagate through the droplet and the number of electrons N_{eh} which can overcome the electrostatic barrier at the rear side can be estimated by [25]:

$$N_{eh} \approx \frac{m_e c^2 4\pi\varepsilon_0 r_L}{e^2}(\gamma - 1),\tag{11.2}$$

where r_L—scales roughly with the laser beam radius which produces an electron bunch with a similar radius. For the experiment with $I_{18} = 2$ one obtains roughly $T_h \approx 190$ keV and $N_{eh} \approx 5 \times 10^8$ for $r_L = 5$–8 μm which can account for a target charge of about 0.10–0.17 nC. The charge is increasing on a picosecond time scale while the hot electrons are leaving. When no more electrons are able to escape the charge is decreasing due to charge balancing of surrounding electrons from the adjacent droplets and the background gas. The electric field of the charge can field-ionize the neighbor droplets as well as the background gas which is arising from the evaporation of the droplets. Also the spatial wings of the laser field can ionize the background gas (see Sect. 11.2). Hence, the droplets are surrounded by a relatively large charge reservoir which can account for charge compensation.

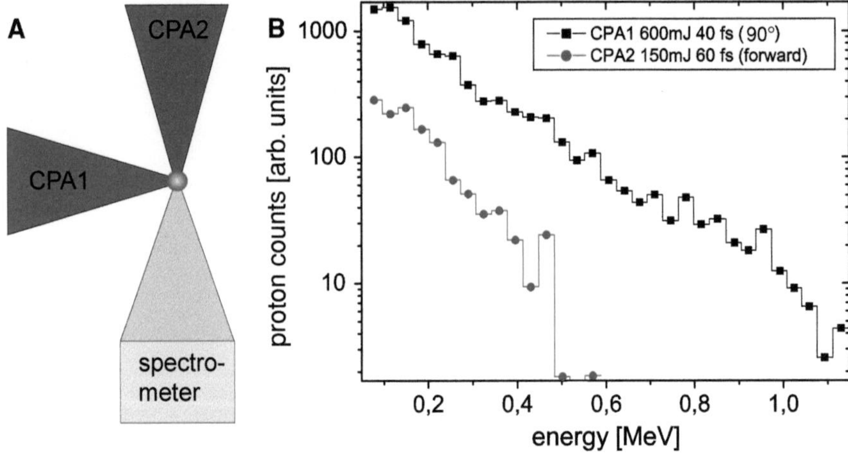

Fig. 11.9 **a** Setup of the spectra measurements. **b** Proton signal emitted from the irradiated droplets in laser (CPA2) direction (*circles*). The *squares* represents the proton signal when irradiating the droplets with CPA1

Therefore the charge is decreasing in time although the droplet is not grounded directly.

In the central part of the images in Fig. 11.8 an undistorted mesh structure is still visible. This is a result of the time integration over 65 ps. The charge is compensated at a time scale of several tens of picoseconds. Protons which arrive at the droplet later are no longer deflected by the field.

Additionally, at the target rear side in direction of the laser propagation, the mesh structure is blurred. This asymmetric deflection is better visible if the mesh is removed. In Fig. 11.8 the pictures of the lower row show the same behavior, a radially symmetric deflection and a broad channel-like structure. This structure in the proton images is characterized by a lower proton number and the smearing of the mesh structure. This is caused by a propagating ion front, time integrated over several picoseconds. The velocity of the ion front can be estimated from the proton spectra emitted from the droplets which was measured simultaneously. The maximum energy of the protons was about 0.5 MeV which corresponds to a velocity of $8 \cdot 10^{6}$ m/s (cf. Fig. 11.9).

As suggested in reference [26, 27] the electric field has a plateau like region between target and front and a strong enhancement directly at the front due to charge separation (see Sect. 3.3). Nevertheless, the integration over a time period of several tens of picoseconds prevents a direct observation of the exact field structure.

To explain the directional ion emission a 2D-PIC simulation [28] was applied by T. Toncian in the scope of the TR18 program.[1] In Fig. 11.10 the distribution of

---

[1] DFG- –Sonderforschungsbereich Transregio (TR18) http://www.tr18.de/.

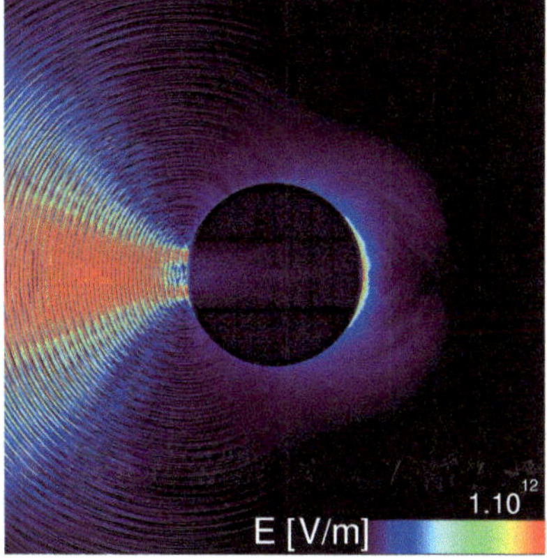

Fig. 11.10 Distribution of
the electric field at $t = 220\,\text{fs}$
as a result from a 2D-PIC
simulation (laser from the left)

the electric field is shown at $t = 220$ fs. A strong enhancement at the rear side of
the droplet (to the right) is visible. This enhancement is caused by electrons which
spread around the droplet, and leads to a directional ion emission.

The calculation predicts such an enhancement only if the laser pulse hits the
droplet centrally (see also [12]). Thus, the difference to the images of Fig. 11.6 can
be explained where the droplets are maybe hit slightly off-axis.

11.4 Three-dimensional Particle Tracing

For a quantitative analysis simulations have been carried out with a commercial
3D particle tracer (http://www.pulsar.nl/gpt). With the help of the tracer the
deflection of the proton probe beam caused by electric fields can be calculated in

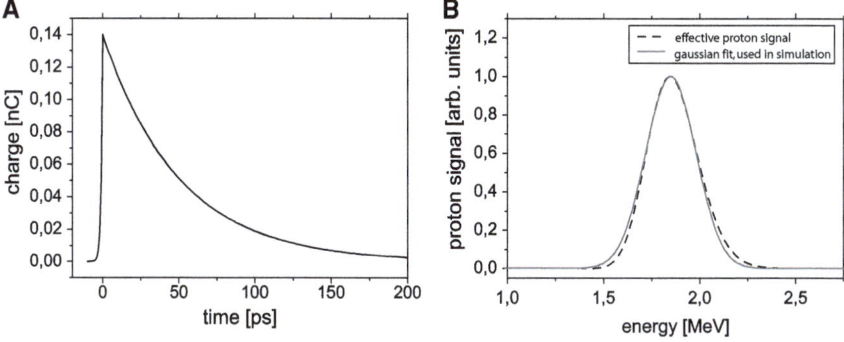

Fig. 11.11 **a** Temporal evolution of the droplet charge ($\Delta t_1 = 1$ ps, $\Delta t_2 = 50$ ps). **b** Assumed
proton spectrum used in the simulations (*solid line*) and the effective proton signal which was
detected in the experiment (*dashed line*). The narrow energy width and the shape is caused by the
gating of the MCP

three dimensions. The energy interval of the detected protons was taken into account (cf. Fig. 11.11b). The positive charge of the droplet was assumed as a point charge which increases and decreases exponentially in time (cf. Fig. 11.11a).

$$Q = Q_0 \cdot \begin{cases} \exp\left(\frac{t}{\Delta t_1}\right) & \text{if } t < 0 \\ \exp\left(-\frac{t}{\Delta t_2}\right) & \text{if } t > 0 \end{cases} \tag{11.3}$$

By comparing the tracing results and the experimental measurements the evolution of the charge could be estimated. The maximum charge was assumed to be 0.14 nC which fits very well with the analytical estimations. Also a minimum time constant for the decreasing field could be determined. The field increases exponentially in only a few picoseconds ($\Delta t_1 = 1$ ps). The exponential decrease in the simulation was assumed to be ($\Delta t_2 = 50$ ps). A faster charge compensation would not reproduce the recorded pictures. The radial deflection—the ring like structure—would not be visible. Also a much slower decrease would be in contradiction to the recorded images at different times. The deflection is only visible in a small time window—similar to the integration time of about 65 ps which fits with the results from the particle tracing.

The result of the simulation, shown in Fig. 11.12b, reproduces the experimental observations very well. The protons are deflected radially symmetrical where the mesh structure shows deformations. Also the reticle like structure of the central horizontal and vertical spacing is visible in the simulated picture. In the particle tracing 10^6 protons where considered for the simulation. Differences in the pictures are visible at the rear side (right-hand side) of the water droplet, whereas the laser comes from the left-hand side. In the experimental observations a distortion and blurring of the mesh structure arises. As discussed above this is caused by the moving ion front in laser forward direction.

In Fig. 11.13 a directed ion front was simulated additionally. For this a rough assumption was made: The moving ion front is simulated by a moving point charge. It was assumed that the field between the moving front and the droplet is zero. This is due to the fact that only a homogenous field between front and target exists which decreases very fast ($\sim t^{-2}$) [27, 29]. The adjacent droplets are field ionized by the field of the charged central droplet and ionized by the wings of the laser beam. The positively charged central droplet leads to a positive charge of the

Fig. 11.12 a Proton image with inserted mesh. **b** Simulation with 1×10^6 particles

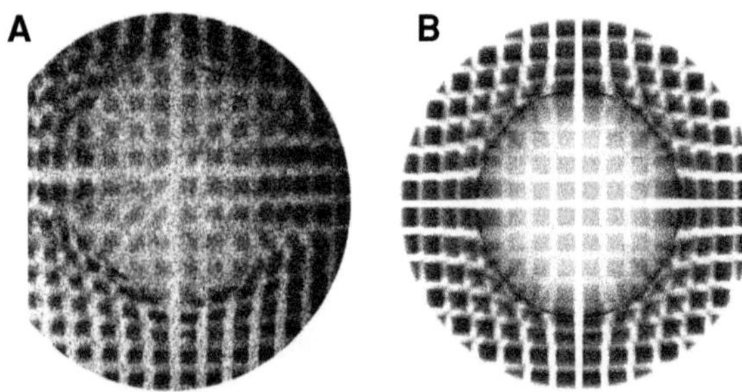

Fig. 11.13 a Proton
image with a pronounced
asymmetric deflection.
b Result of a simulation
done with the particle tracer
(laser from the left-hand side)

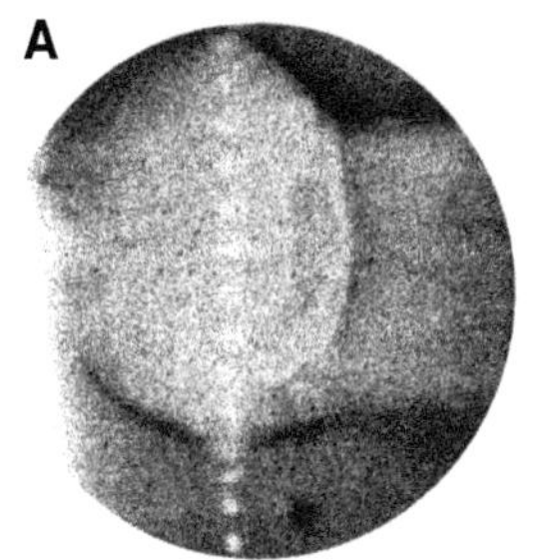

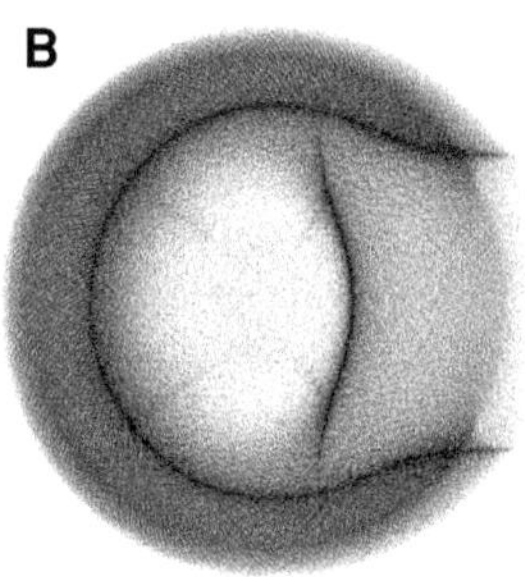

two adjacent droplets. This effect was already seen in reference [30] where a metal wire was irradiated by an intense laser pulse and a proton emission from a second wire placed 250 μm away was observed and explained by the charge up effect. Thus, it was assumed in the simulation than the neighbor droplets are positively charged by 1/6 of the overall charge. The result of the simulation, shown in Fig. 11.13 visualizes the creation of the channel-like structure by a moving localized field—like the field of the expanding ion front (see Chap. 12).

In summary, laser irradiated water-droplets were investigated with proton imaging. A positive charge ($\sim$0.14 nC) of the droplets caused by the escaped precursor electrons and a directional ion emission in laser forward direction with energies up to 0.5 MeV were observed. Simulations of the deflection with a 3D particle tracer are in a good agreement with the experimental results.

References

1. R.A. Snavely, M.H. Key, S.P. Hatchett, T.E. Cowan, M. Roth, T.W. Phillips, M.A. Stoyer, E.A. Henry, T.C. Sangster, M.S. Singh, S.C. Wilks, A. MacKinnon, A. Offenberger, D.M. Pennington, K. Yasuike, A.B. Langdon, B.F. Lasinski, J. Johnson, M.D. Perry, E.M. Campbell, Intense high-energy proton beams from Petawatt-laser irradiation of solids. Phys. Rev. Lett. **85**(14), 2945 (2000)
2. J. Badziak, Laser-driven generation of fast particles. Opto-Electron. Rev. **15**(1), 1–12 (2007)
3. T. Okada, A.A. Andreev, Y. Mikado, K. Okubo, Energetic proton acceleration and bunch generation by ultraintense laser pulses on the surface of thin plasma targets. Phys. Rev. E **74**(2), 5 (2006)
4. S.C. Wilks, A.B. Langdon, T.E. Cowan, M. Roth, M. Singh, S. Hatchett, M.H. Key, D. Pennington, A. MacKinnon, R.A. Snavely, Energetic proton generation in ultra-intense laser–solid interactions. Phys. Plasmas **8**(2), 542–549 (2001)
5. P.K. Patel, A.J. Mackinnon, M.H. Key, T.E. Cowan, M.E. Foord, M. Allen, D.F. Price, H. Ruhl, P.T. Springer, R. Stephens, Isochoric heating of solid-density matter with an ultrafast proton beam. Phys. Rev. Lett. **91**(12), 125004 (2003)
6. B.M. Hegelich, B.J. Albright, J. Cobble, K. Flippo, S. Letzring, M. Paffett, H. Ruhl, J. Schreiber, R.K. Schulze, J.C. Fernandez, Laser acceleration of quasi-monoenergetic MeV ion beams. Nature **439**(7075), 441–444 (2006)
7. H. Schwoerer, S. Pfotenhauer, O. Jackel, K.U. Amthor, B. Liesfeld, W. Ziegler, R. Sauerbrey, K.W.D. Ledingham, T. Esirkepov, Laser-plasma acceleration of quasi-monoenergetic protons from microstructured targets. Nature **439**(7075), 445–448 (2006)

8. S. Ter-Avetisyan, M. Schnurer, P.V. Nickles, M. Kalashnikov, E. Risse, T. Sokollik, W. Sandner, A. Andreev, V. Tikhonchuk, Quasimonoenergetic deuteron bursts produced by ultraintense laser pulses. Phys. Rev. Lett. **96**(14), 145006 (2006)
9. A.V. Brantov, V.T. Tikhonchuk, O. Klimo, D.V. Romanov, S. Ter- Avetisyan, M. Schnurer, T. Sokollik, P. V. Nickles, Quasi-mono energetic ion acceleration from a homogeneous composite target by an intense laser pulse. Phys. Plasmas **13**(12), 10 (2006)
10. M. Schnurer, S. Ter-Avetisyan, S. Busch, E. Risse, M.P. Kalachnikov, W. Sandner, P. Nickles, Ion acceleration with ultrafast laser driven water droplets. Laser Part. Beams **23**(3), 337–343 (2005)
11. J. Psikal, J. Limpouch, S. Kawata, A.A. Andreev, PIC simulations of femtosecond interactions with mass-limited targets. Czechoslov. J. Phys. **56**, B515–B521 (2006)
12. A.J. Kemp, H. Ruhl, Multispecies ion acceleration off laser-irradiated water droplets. Phys. Plasmas **12**(3), 033105–033110 (2005)
13. S. Karsch, S. Düsterer, H. Schwoerer, F. Ewald, D. Habs, M. Hegelich, G. Pretzler, A. Pukhov, K. Witte, R. Sauerbrey, High-intensity laser induced ion acceleration from heavy-water droplets. Phys. Rev. Lett. **91**(1), 015001 (2003)
14. S. Busch, M. Schnurer, M. Kalashnikov, H. Schonnagel, H. Stiel, P.V. Nickles, W. Sandner, S. Ter-Avetisyan, V. Karpov, U. Vogt, Ion acceleration with ultrafast lasers. Appl. Phys. Lett. **82**(19), 3354–3356 (2003)
15. C. Iaconis, I.A. Walmsley, Spectral phase interferometry for direct electric-field reconstruction of ultrashort optical pulses. Opt. Lett. **23**(10), 792–794 (1998)
16. S. Ter-Avetisyan, M. Schnurer, P.V. Nickles. Time resolved corpuscular diagnostics of plasmas produced with high-intensity femtosecond laser pulses. J. Phys. D-Appl. Phys. **38**(6), 863–867 (2005)
17. J. Eggers, Nonlinear dynamics and breakup of free-surface flows. Rev. Mod. Phys. **69**(3), 865 (1997)
18. S. Düsterer, Laser–plasma interaction in droplet-targets. Ph.D. thesis, Friedrich-Schiller-Universität Jena, 2002
19. T. Pfeifer, Adaptive control of coherent soft X-rays. Ph.D. thesis, Bayerischen Julius-Maximilians-Universität Würzburg, 2004
20. M. Borghesi, S. Kar, L. Romagnani, T. Toncian, P. Antici, P. Audebert, E. Brambrink, F. Ceccherini, C.A. Cecchetti, J. Fuchs, M. Galimberti, L.A. Gizzi, T. Grismayer, T. Lyseikina, R. Jung, A. Macchi, P. Mora, J. Osterholtz, A. Schiavi, O. Willi, Impulsive electric fields driven by high-intensity laser matter interactions. Laser Part. Beams **25**(1), 161–167 (2007)
21. P. McKenna, D.C. Carroll, R.J. Clarke, R.G. Evans, K.W.D. Ledingham, F. Lindau, O. Lundh, T. McCanny, D. Neely, A.P.L. Robinson, L. Robson, P.T. Simpson, C.G. Wahlstrom, M. Zepf, Lateral electron transport in high-intensity laser-irradiated foils diagnosed by ion emission. Phys. Rev. Lett. **98**(14), 145001 (2007)
22. M.G. Haines, Thermal instability and magnetic field generated by large heat flow in a plasma, especially under laser-fusion conditions. Phys. Rev. Lett. **47**(13), 917 (1981)
23. M. Borghesi, L. Romagnani, A. Schiavi, D.H. Campbell, M.G. Haines, O. Willi, A.J. Mackinnon, M. Galimberti, L. Gizzi, R.J. Clarke, S. Hawkes, Measurement of highly transient electrical charging following high-intensity laser-solid interaction. Appl. Phys. Lett. **82**(10), 1529–1531 (2003)
24. J. Fuchs, P. Antici, E. d'Humieres, E. Lefebvre, M. Borghesi, E. Brambrink, C.A. Cecchetti, M. Kaluza, V. Malka, M. Manclossi, S. Meyroneinc, P. Mora, J. Schreiber, T. Toncian, H. Pepin, R. Audebert, Laser-driven proton scaling laws and new paths towards energy increase. Nat. Phys. **2**(1), 48–54 (2006)
25. A.A. Andreev, T. Okada, K.Y. Platonov, S. Toraya, Parameters of a fast ion jet generated by an intense ultrashort laser pulse on an inhomogeneous plasma foil. Laser Part. Beams **22**(4), 431–438 (2004)
26. P. Mora, Thin-foil expansion into a vacuum. Phys. Rev. E **72**(5), 5 (2005)
27. P. Mora, Plasma expansion into a vacuum. Phys. Rev. Lett. **90**(18), 185002 (2003)

28. G. PIC-code by H. Ruhl, Ruhr University Bochum
29. L. Romagnani, J. Fuchs, M. Borghesi, P. Antici, P. Audebert, F. Ceccherini, T. Cowan, T. Grismayer, S. Kar, A. Macchi, P. Mora, G. Pretzler, A. Schiavi, T. Toncian, O. Willi, Dynamics of electric fields driving the laser acceleration of multi-MeV protons. Phys. Rev. Lett. **95**(19), 195001 (2005)
30. F.N. Beg, M.S. Wei, A.E. Dangor, A. Gopal, M. Tatarakis, K. Krushelnick, P. Gibbon, E.L. Clark, R.G. Evans, K.L. Lancaster, P.A. Norreys, K.W.D. Ledingham, P. McKenna, M. Zepf, Target charging effects on proton acceleration during high-intensity short-pulse laser–solid interactions. Appl. Phys. Lett. **84**(15), 2766–2768 (2004)

Chapter 12
Streak Deflectometry

In the previous chapters proton deflectometry was used to investigate electric fields. Therefore two-dimensional temporal snapshots of the probed plasma were applied. But they always deliver temporally integrated images of the transient fields at different probing times. In future experiments the newly applied gated MCP technique can be improved to generate short time integrated snapshots with a temporal resolution of a few picoseconds (cf. Appendix B). New high voltage pulse generators are available, optimized to gate multi-channel plates with a time window of 5 ns down to hundreds of picoseconds. Nevertheless a series of many snapshots is required to reconstruct the temporal evolution of the fields in detail.

In the following chapter a novel method will be discussed which allows a continuous record of the field evolution for the first time. This method is named "streak deflectometry". The proton probe beam is streaked dependent on the proton energy in one dimension due to a simple magnet. Thus, the deflection in the probed field of every proton energy can be traced back to a continuous temporal evolution of the field.

The "streak deflectometry" method was used to investigate the rear side of a laser irradiated thin foil. Recent experiments (e.g. [1]) show the necessity to investigate extended electric fields which exist much longer than the laser pulse duration. The measurements show the field evolution in a time window of several nanoseconds with a resolution of ~ 30 ps. Due to numerous data points a detailed reconstruction of the fields are possible.

12.1 The Proton Streak Camera

If the proton beam is detected with a velocity dispersive detector (e.g. a magnet spectrometer) the influence of electric fields to the protons can be traced back in

T. Sokollik, *Investigations of Field Dynamics in Laser Plasmas with Proton Imaging,* Springer Theses, DOI: 10.1007/978-3-642-15040-1_12,
© Springer-Verlag Berlin Heidelberg 2011

time. Thus, the velocity dependent detection of such a proton beam constitutes itself as a "proton streak camera".

The setup is similar to the one in Chap. 10. The proton beam has been produced by high intensity ($\sim 10^{19}$ W/cm^2) 40 fs laser irradiation (CPA1) of 12 μm thick aluminium foils. A second laser (CPA2) was used to produce a second plasma which is to be probed. This Nd:glass laser provides 1.5 ps long laser pulses at a peak power of about 5 TW (Chap. 5), and is synchronized to the CPA1—Ti:Sa laser with a precision of a few picoseconds. The interaction target was a 12 μm thick, curved aluminium foil, which was mounted as a stripe of about 8–10 mm width and bent with a radius of about 5 mm. The CPA2—laser irradiated the concave side and the proton beam probed the rear of the target at 90° to the target normal at the CPA2 interaction point.

Additionally to the conventional setup a thin slit and a magnet are added (Fig. 12.1). This setup acts like a spectrometer—the beam is deflected depending on the proton velocity. In direction of the ~ 200 μm slit the deflections due to the probed electric fields are measurable. In order to trace the proton deflection perpendicular to the dispersion direction, the proton beam is intersected by a mesh. Each stripe corresponds to a small part of the proton beam which passes the interaction target at a defined distance and its deflection is projected to the detector. In fact, for every proton energy the deflection is recorded. Since each proton energy corresponds to a different probing time (arrival time at the object to probe) the continuous temporal evolution of the fields is recorded.

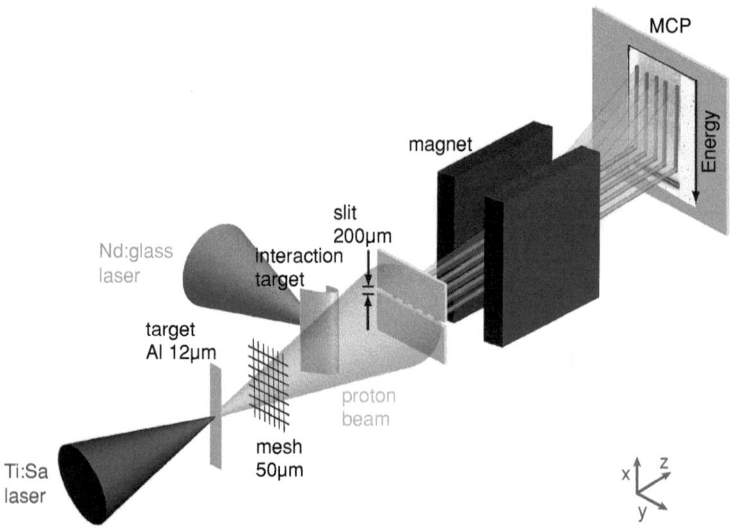

Fig. 12.1 Experimental setup for proton streak measurements. The proton beam emerges from the rear side of a Ti:Sa laser irradiated ($\sim 10^{19}$ W/cm^2) foil. A mesh is placed in the proton beam at 30 mm distance from source and the proton beam probes the rear side of a laser irradiated (10^{17}-10^{18} W/cm^2) 2nd target foil. The proton beam propagates through the magnetic field which deflects the protons dependent on their velocity

The time resolution given by the energy resolution of the magnetic spectrometer and the length of the proton flight path from the 1st (proton producing) to the 2nd (interaction) target is about 30 ps.

12.2 Streaking Transient Electric Fields

An undisturbed image (without the incident of CPA2) recorded with the "streak deflectometry" setup is shown in Fig. 12.2a. Along the x-axis protons are dispersed according to their velocity which determines the arrival time of the probe

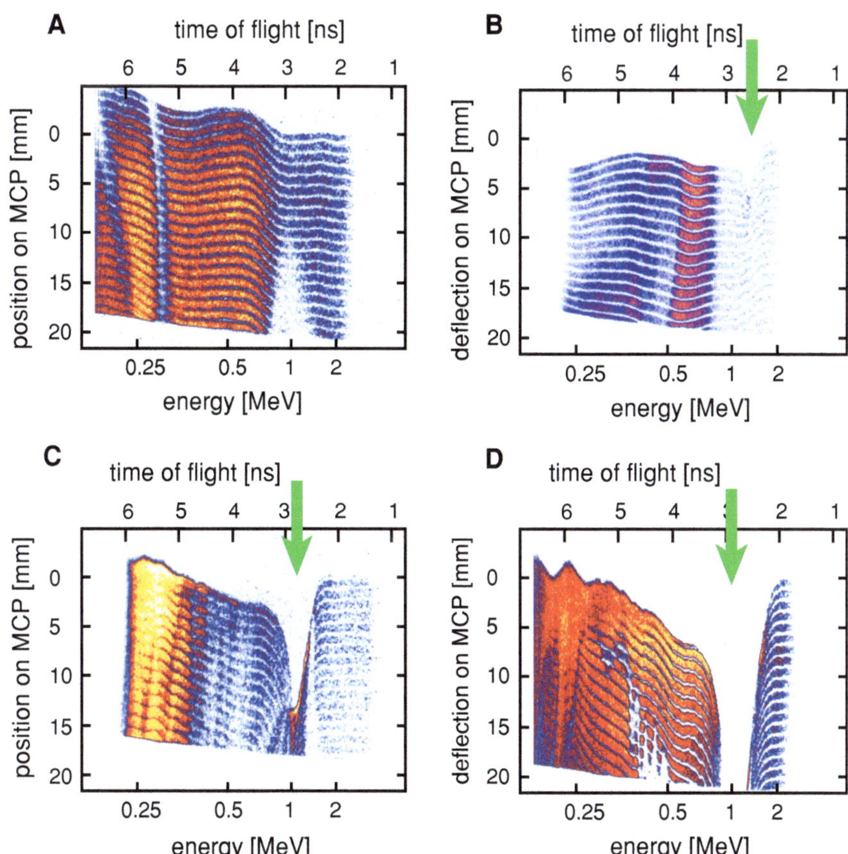

Fig. 12.2 Proton streak images: Position against energy. The latter corresponds to the time of flight to the interaction target (upper abscissa). The *arrow* indicates CPA2 laser irradiation on the interaction target, the edge of the interaction target is at the upper boundary of the traces; **a** no irradiation of the interaction target, **b** 0.7 J pulse energy of CPA2, **c** 2.3 J and **d** 5 J

pulse at the interaction target (upper abscissa). The y-axis shows the position at the detector. Even without the incident laser (CPA2) the pointing of the proton beam causes fluctuations at energies below 0.8 MeV. This is based on small movements of the proton source during emission which causes a "wiggly" trace as determined in reference [2, 3] (see also Chap. 8 and Sect. 6.3). This "wiggling" is fluctuating from shot to shot. Therefore variation of the beam pointing has been calculated from ten shots for data analysis (cf. Fig. 12.4b).

Figure 12.2b–d shows the proton deflections when the CPA2 pulse irradiates the interaction target at intensities of $(1–9) \times 10^{17}$ W/cm^2 (0.7–5 J pulse energy). The energy cut-off of the protons sets the maximum energy, whereas the minimum energy in the measurements is set by the MCP gating or by the boundary of the detector.

High energetic protons reach the interaction target at first. At this time no fields have been created due to the absence of the second laser pulse (CPA2). As the laser pulse hits the target, electric fields are created and thus the proton probe beam is deflected away from the target surface. At first, the deflection increases and decreases relatively fast within a few picoseconds. Note that the recorded traces show a convolution with the field extension. The deflections strongly depend on the distance to the target surface. Protons passing closer to the interaction target are deflected stronger than protons being far away. At later times the deflection decreases slowly (up to 5 ns), but now the deflection does not depend on the distance to the target surface.

This observation can be explained by the presence of two electric fields. One field (Field 1) is created due to the charge distribution at the ion front. The electric field peaks directly at the ion front [4] which expands with an assumed velocity of about 10^6–10^7 m/s. The field is strongly localized and decreases in time $(\propto t^{-1})$ [4, 5] and is responsible for the fast and strong increasing and decreasing deflection.

The second field (Field 2) is created due to leaving precursor electrons. The target gets positively charged. The deflection (Y) of the proton probe beam, propagating in z-direction, can be estimated as follows:

The electric field of a charged cylinder is determined by:

$$E_{el} = \frac{\sigma}{\varepsilon_0} \frac{R}{r}, \qquad r > R \tag{12.1}$$

where σ is the charge density, R the radius of the cylinder and r the distance to the cylinder axis. The y-component of the electric field can be written as:

$$E_{el,y} = \frac{\sigma}{\varepsilon_0} \frac{R}{r} \cdot \frac{y}{r}, \qquad r = \sqrt{z^2 + y^2}, \quad E_{el,y} = E_{el} \cdot \frac{y}{r}. \tag{12.2}$$

Integrating over the electric field along the path of the protons (assuming small deflections) leads to the velocity v_y of the protons in y-direction.

$$E_{el,y} \cdot e = m_p \cdot \frac{d^2 y}{dt^2} \tag{12.3}$$

$$v_y = \int \frac{E_{el,y} \cdot e}{m_p \cdot v_z} \, dz, \qquad v_z = \frac{dz}{dt} \tag{12.4}$$

$$v_y = \frac{\sigma \cdot \pi \cdot R \cdot e}{\varepsilon_0 \cdot m_p \cdot v_z} \tag{12.5}$$

The ratio of the velocities (v_y/v_z) and the distance to the detector L determine the position of the protons on the detector.

$$Y = \frac{\sigma \cdot \pi \cdot R \cdot e}{\varepsilon_0 \cdot m_p \cdot v_z^2} \cdot L, \qquad Y = \frac{v_y}{v_z} \cdot L \tag{12.6}$$

Protons are deflected regardless to their incoming original position on y-axis (y) by a similar value [6]. This is visible at the low energetic part of the experimental traces.

For a quantitative analysis of Fig. 12.2c, with a 2.3 J incident laser pulse, a model was used on the basis of [5] with an additional charge of the cylinder.

12.3 Fitting Calculations

In the following the measurement shown in Fig. 12.2c is analyzed in detail (CPA2—2.3 J). The deflection of Field 2 was calculated similar to Eq. 12.6. The field relates to the electric field of a charged cylindrical surface with a total charge of 2×10^{-9} C extending over 8 mm along the target surface. It increases in time with an exponential coefficient of 0.135 ns and decays at about 1 ns.

The field at the ion front (Field 1) was assumed to increase and decrease exponentially with the distance to the ion front (scale length: 100 μm). The ion front expands with about 10^7 m/s while the field strength decreases with the cube of the distance to the target surface. The maximum strength of Field 1 at the surface is about 4×10^8 V/m. The field starts with an extension of 5 mm along the target surface. In Fig. 12.3 the structure of Field 1 (the expanding ion front) is shown schematically for one special time.

In Fig. 12.4 the calculated field components together with the experimental traces of Fig. 12.2c are plotted. In Fig. 12.2b also the fluctuations of the proton beam pointing, estimated over ten shots without incident laser, were taken into account. The determination of the strength of Fields 1 and 2 is subjected to an error of about 7 and 20%, respectively.

The combination of both fields describes the whole recorded experimental picture very well, although some assumptions have to be considered more precisely.

For example the large extension of the fields are necessary for the width of the deflection peak. A larger or smaller extension would not fit the experimental traces. Additionally this fact is confirmed by the two-dimensional proton images in Chap. 10 and the source characterization measurements in Chap. 8.

Fig. 12.3 Schematical structure of Field 1 at the beginning of the expansion

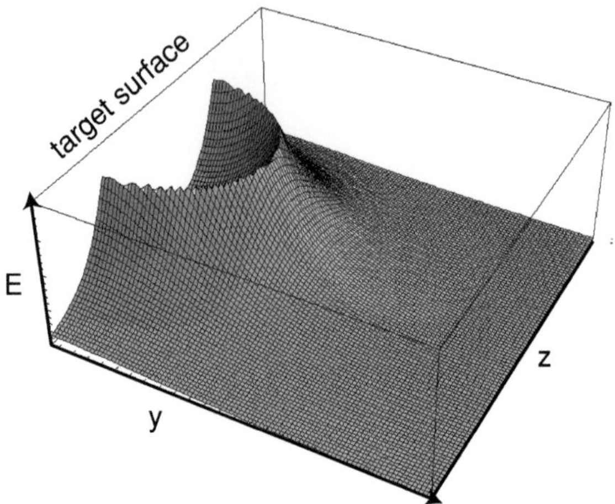

Also the estimated expansion velocity of 10^7 m/s is a result of the fitting to the experimental traces. Lower velocities would result in a too narrow deflection curve and non-coincident maxima of the different traces. In the simultaneously measured spectra, the maximum energy of the protons emitted from the interaction target was determined to be less than 0.5 MeV which corresponds to a velocity of 10^7 m/s. Thus, the calculation delivers an overestimated velocity of the front.

The estimated temporal evolution of Field 2 shows a relatively slow increase of about 135 ps. One would expect an increase of the field within a few picoseconds when the precursor electrons have left the target. In fact the lateral spread of the electrons over the target surface has not been taken into account in the calculations. Also shielding effects by the expanding hot electron cloud are not considered. Hence the estimated time constant of the increasing field represents the upper limit. However, the decay time of the field is less influenced by these effects. Thus, the estimated time constant delivers a representative value.

The estimated maximum electric field strength at the beginning of the expansion process is plotted in Fig. 12.5 for different Nd:glass laser energies. This field strength represents a temporal averaged value of the acceleration field and the field of the expansion front. A nearly exponential dependence of the maximum field and the laser energy is indicated.

Regarding to the scaling laws given in reference [7] for the TNSA-process and the laser energy conversion [8] one would expect a dependence of the electric field ($E_{0,el}$) as follows:

$$E_{0,el} \propto \sqrt{E_{\text{laser}}}, \tag{12.7}$$

where E_{laser} is the laser energy. In Chap. 10 this behavior has been observed at the same experimental conditions but at higher pulse energies (CPA2). This could be an indication that laser intensities of $< 5 \times 10^{17}$ W/cm^2 give a lower threshold of

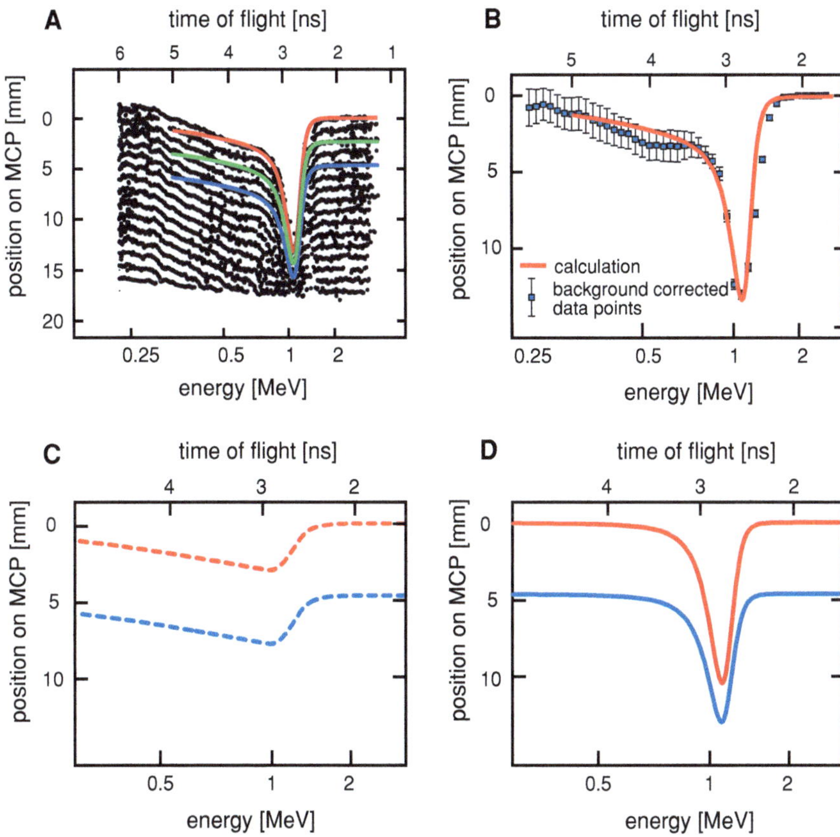

Fig. 12.4 a Coordinate read out of traces together with calculated deflection (cf. text), **b** comparison of the superimposed field fit to the measured proton beam deflection. Error bars are caused by the shot to shot fluctuation of the proton beam pointing (average of 10 shots), **c** deflection by Field 2—the "target charge up" field alone, **d** deflection by Field 1—a moving field front coupled to an ion front

the analytical estimations of the TNSA-process at this target and laser parameter configuration.

12.4 Particle Tracing

In fact, the model described above is oversimplifying the physical processes. Thus, a second model was developed taking into account the physical processes more precisely. The calculation, including a particle tracing, was done by M. Amin (University of Düsseldorf) in the scope of cooperation in the TR18 program. In this approach Field 1 was constructed by modelling an one-dimensional plasma

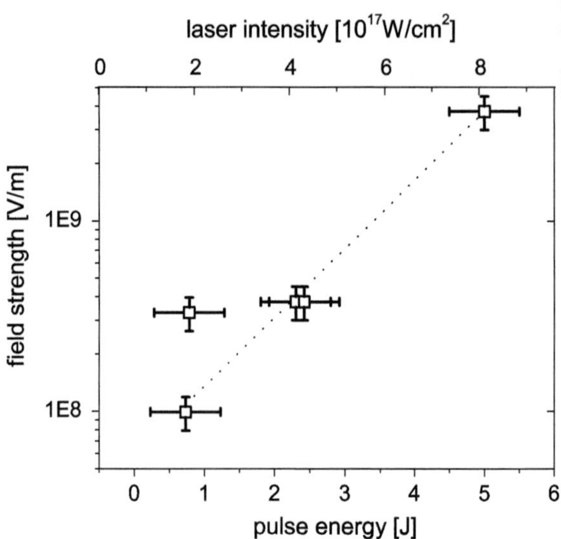

Fig. 12.5 Maximum field strength for different pulse energies of CPA2

expansion into vacuum according to [4, 5]. Spatially, the electric field shows a plateau region which is followed by an exponential rise up to the peak at the front and then decays as $(1 + r/l)^{-1}$ where l is the field scale length and r is the distance to the target surface. The field front moves away from the target surface while the electric field in the plateau region decays as $(1 + t/\tau)^{-2}$ whereas at the peak and the subsequent region the field decays as $(1 + t/\tau)^{-1}$ (τ is the decay constant). Field 2 is the field of a charged cylinder, shielded by the charge of the field front. The two fields are superposed and applied to field lines normal to the target surface.

The electrons involved in the plasma expansion, assumed by the scaling law given in reference [9] with a temperature of roughly 100 keV and carrying about 5–10% of the focused laser energy [10], spread over the rear side of the target with a Gaussian density distribution of about 6 mm FWHM.[1] The field scale length l was supposed to be 100 μm, similar to reference [5]. The following parameters could be fitted also to the experimental data: the decay constant τ (3 ps) of Field 1, the front propagation velocity (10^6 m/s), the maximum charge density (Field 2) on the target surface (10^{-4} C/m^2), the linear grow within 10 ps and the exponential decay time (600 ps) of the target charge.

Figure 12.6a shows a simulated proton streak measurement including an extracted deflection curve from the experimental data. By varying the decay time in the simulation, the shape of the deflection peak could be influenced and thus the decay time could be fitted. The wavy structure in vicinity of the deflection peak can be seen both in the experimental and in the simulated picture which was

---

[1] Similar to the calculations in the previous section, this assumption has to be made to reproduce the width of the deflection curve.

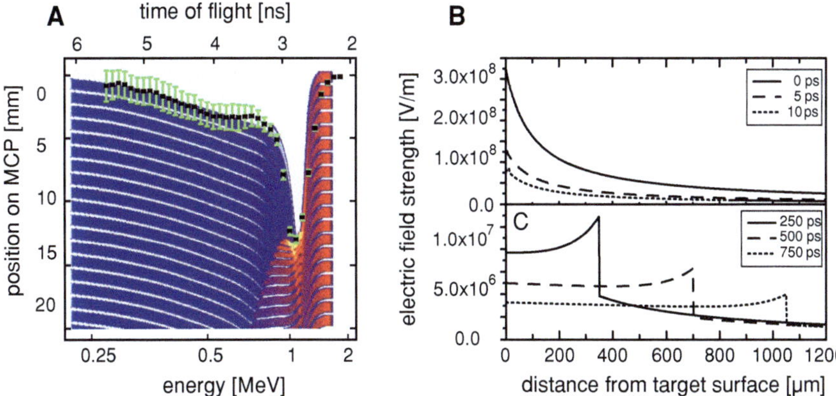

Fig. 12.6 a Simulated proton streak measurement including an extracted deflection curve from the experimental data. **b, c** Electric field strength against distance from the target surface for different times. **b** shows the electric field evolution shortly after the incidence of CPA2. The graphs correspond to 0 ps (*solid line*), 5 ps (*dotted line*), 10 ps (*dashed line*) after the incidence of the peak of the pulse. **c** shows the long-run evolution: 250 ps (*solid line*), 500 ps (*dotted line*), 750 ps (*dashed line*). Simulation done by M. Amin—University of Düsseldorf

matched by varying the front propagation velocity. This structure is caused by the interaction of the probe particles with the field front. Figure 12.6 reproduces qualitatively the measured deflection around the peak and the behavior at the trailing edge.

The applied electric fields are plotted in Fig. 12.6b, c. The electric field peaks at $t = 0$ at the target surface at about 3×10^8 V/m. It shows a sharp decrease during the first picoseconds but then persists on a lower level while the front propagates through the experimentally observable region.

With the analytical calculations and the particle tracing similar values for the maximum electric field strength, about $(3\text{–}4) \times 10^8$ V/m, of Field 1 were estimated. Also the background charge and its decay time was estimated within the same order of magnitude about $(0.3\text{–}1) \times 10^{-4}$ C/m^2 within $(0.6\text{–}1)$ ps.

In summary the novel imaging method, called proton "streak deflectometry", allows measurements of real-time dynamics of transient and intense fields in laser plasma interactions. In particular the fields on the rear of a laser irradiated metal foil which are responsible for the process of laser proton acceleration were investigated. The observed streak images were qualitatively explained by the temporal and 1D-spatial development of two electric fields arising from charge-up and charge compensation at a nanosecond timescale, and ion front propagation at a timescale of several hundreds of picoseconds. Thus, effects of energetic electron generation and extended lateral transport which leads to transient electric fields ($\sim 10^8$ V/m) with millimeter lateral extension was observed.

References

1. T. Toncian, M. Borghesi, J. Fuchs, E. d'Humieres, P. Antici, P. Audebert, E. Brambrink, C.A. Cecchetti, A. Pipahl, L. Romagnani, O. Willi, Ultrafast laser-driven microlens to focus and energy-select mega-electron volt protons. Science **312**(5772), 410–413 (2006)
2. J. Schreiber, S. Ter Avetisyan, E. Risse, M.P. Kalachnikov, P.V. Nickles, W. Sandner, U. Schramm, D. Habs, J. Witte, M. Schnurer, Pointing of laser-accelerated proton beams. Phys. Plasmas **13**(3), 033111 (2006)
3. S. Ter-Avetisyan, M. Schnurer, T. Sokollik, P.V. Nickles, W. Sandner, H.R. Reiss, J. Stein, D. Habs, T. Nakamura, K. Mima, Proton acceleration in the electrostatic sheaths of hot electrons governed by strongly relativistic laser-absorption processes. Phys. Rev. E **77**(1), 016403 (2008)
4. P. Mora, Plasma expansion into a vacuum. Phys. Rev. Lett. **90**(18), 185002 (2003)
5. L. Romagnani, J. Fuchs, M. Borghesi, P. Antici, P. Audebert, F. Ceccherini, T. Cowan, T. Grismayer, S. Kar, A. Macchi, P. Mora, G. Pretzler, A. Schiavi, T. Toncian, O. Willi, Dynamics of electric fields driving the laser acceleration of multi-MeV protons. Phys. Rev. Lett. **95**(19), 195001 (2005)
6. C. Gerthsen, H. Vogel, *Physik*, vol. 17. Springer-Verlag, Berlin (1993)
7. S.P. Hatchett, C.G. Brown, T.E. Cowan, E.A. Henry, J.S. Johnson, M.H. Key, J.A. Koch, A.B. Langdon, B.F. Lasinski, R.W. Lee, A.J. Mackinnon, D.M. Pennington, M.D. Perry, T.W. Phillips, M. Roth, T.C. Sangster, M.S. Singh, R.A. Snavely, M.A. Stoyer, S.C. Wilks, K. Yasuike, Electron, photon, and ion beams from the relativistic interaction of Petawatt laser pulses with solid targets. Phys. Plasmas **7**(5), 2076–2082 (2000)
8. S. Busch, M. Schnurer, M. Kalashnikov, H. Schonnagel, H. Stiel, P.V. Nickles, W. Sandner, S. Ter-Avetisyan, V. Karpov, U. Vogt, Ion acceleration with ultrafast lasers. Appl. Phys. Lett. **82**(19), 3354–3356 (2003)
9. J. Fuchs, P. Antici, E. d'Humieres, E. Lefebvre, M. Borghesi, E. Brambrink, C.A. Cecchetti, M. Kaluza, V. Malka, M. Manclossi, S. Meyroneinc, P. Mora, J. Schreiber, T. Toncian, H. Pepin, R. Audebert, Laser-driven proton scaling laws and new paths towards energy increase. Nat. Phys. **2**(1), 48–54 (2006)
10. M. Schnurer, M.P. Kalashnikov, P.V. Nickles, T. Schlegel, W. Sandner, N. Demchenko, R. Nolte, P. Ambrosi. Hard X-ray-emission from intense short-pulse laser plasmas. Phys. Plasmas **2**(8), 3106–3110 (1995)

Chapter 13
Summary and Outlook

Proton acceleration due to the interaction of high intense laser pulses with matter has a high potential for future application. Therefore a more detailed understanding of the physical processes is required to optimize the beam attributes for these purposes. One recently realized application is the proton imaging which was used in the presented thesis to investigate the acceleration processes.

In the frame of this thesis novel imaging techniques were developed and used to study different target systems. The "streak deflectometry" allows a continuous record of electric fields over a nanosecond timescale for the first time. The temporal evolution of these electric fields was studied in detail. They show a large extension and a lifetime of several hundreds of picoseconds and a few picoseconds, respectively. These results help to understand the temporal evolution of the proton source and processes involved in other applications e.g. the micro-lens [1].

For the first time gated MCPs were used for the proton imaging purpose. They allow two-dimensional snapshots of the field evolution. This technique enables the first detailed investigation of irradiated isolated micro-spheres. These mass-limited targets which are electronically not connected to the ground are one of the most promising systems regarding the efficiency and production of tailored proton spectra. As a result of these experiments a field structure was observed and analyzed which can be connected to a directed proton emission. This novel effect could be used to overcome the hindering isotropic proton emission and thus to enhance the usable number of protons for applications. This essential feature in combination with the high repetition rate, quasi-monoenergetic ion beam generation and predictions of efficient conversion of laser to ion energy, increases the attraction of such an ion beam source.

Aside from proton imaging experiments, investigations of the proton beam characteristics have delivered new results concerning the proton source and its temporal evolution. Especially at lower proton energies (< 1 MeV) protons are accelerated from areas of several millimeter extension when irradiating thin foil targets even with focused laser beams of micrometer diameter. It was shown that

T. Sokollik, *Investigations of Field Dynamics in Laser Plasmas with Proton Imaging*, Springer Theses, DOI: 10.1007/978-3-642-15040-1_13,
© Springer-Verlag Berlin Heidelberg 2011

the emission is influenced by the laser parameters especially the temporal contrast of the pulses. Thus, different absorption mechanisms can arise, creating electron currents which cause different proton acceleration areas. These results are partly supported by the imaging experiments and are important for the application of the imaging technique itself. The fluctuation of the proton beam pointing at low energies have to be taken into account for the interpretation of the experimental results.

Future experiments will give a more detailed picture of the acceleration process for mass-limited and isolated targets. The time resolution of the imaging techniques with gated MCPs will be improved further due to new available gating units. One aspired ambitious goal is a trapped single sphere irradiated by the laser pulse and probed by a proton beam. The advantage compared to the water droplet target will be the background-gas free environment. It has been shown that the water droplet can not be regarded as a fully isolated system. The short distance between the droplets and the background-gas caused by evaporation of the droplets, result in a more complex system influencing the emission characteristics. The irradiation of a real isolated sphere with a high contrast and intense laser pulse is a promising goal.

New techniques for temporal pulse cleaning have been developed recently. This evolution enables laser matter interaction with ultra-thin targets of tens of nanometer only. Therewith an enhancement of efficiency of the acceleration process is connected and needs to be investigated in detail. The combination of a high pulse contrast and mass-limited targets could approach a new regime in the particle acceleration.

Reference

1. T. Toncian, M. Borghesi, J. Fuchs, E. d'Humieres, P. Antici, P. Audebert, E. Brambrink, C.A. Cecchetti, A. Pipahl, L. Romagnani, O. Willi, Ultrafast laser-driven microlens to focus and energy-select mega-electron volt protons. Science **312**(5772), 410–413 (2006)

Part IV
Appendix

Chapter 14
Zernike Polynomials

Zernike polynomials are used in Sect. 5.1 to describe the wavefront of the Ti:Sa laser pulse. They are a set of polynomials, defined on the unit circle and consist of an angular function and radial polynomials derived from the Jacobi polynomials. Slightly different definitions exist concerning the normalization. The following formalisms are strongly connected to the definitions used by the wave-front sensor unit and can be found in reference [1]. The polynomials are defined by:

$$\left.\begin{array}{l} Z_{even\,j} = \sqrt{n+1} \cdot R_n^m(r) \cdot \sqrt{2} \cdot \cos m\phi \\ Z_{odd\,j} = \sqrt{n+1} \cdot R_n^m(r) \cdot \sqrt{2} \cdot \sin m\phi \end{array}\right\} \quad m \neq 0 \qquad (14.1)$$

$$Z_j = \sqrt{n+1} \cdot R_n^m(r), \quad m = 0$$

where

$$R_n^m(r) = \sum_{s=0}^{(n-m)/2} \frac{(-1)^s (n-s)!}{s![(n+m)/2 - s]![(n-m)/2 - s]!} r^{n-2s} \qquad (14.2)$$

The ordering index j is a function of n and m. In Fig. 14.1 the used ordering is shown. Another common notation of the polynomials is given by Z_n^m, where negative m indicates *even j*. The first orders can be connected to classical aberration as follows:

$$\begin{array}{lll} Z_2 = Z_1^{-1} & = 2r\cos\phi & \\ Z_3 = Z_1^1 & = 2r\sin\phi & \text{tilt (lateral position)} \\ Z_4 = Z_2^0 & = \sqrt{3}(2r^2 - 1) & \text{defocus (longitudinal)} \\ Z_5 = Z_2^2 & = \sqrt{6}r^2 \sin 2\phi & \\ Z_6 = Z_2^{-2} & = \sqrt{6}r^2 \cos 2\phi & \text{astigmatism (third order)} \\ Z_7 = Z_2^2 & = \sqrt{8}(3r^2 - 2r)\sin\phi & \\ Z_8 = Z_2^{-2} & = \sqrt{8}(3r^2 - 2r)\cos\phi & \text{coma (third order)} \end{array}$$

T. Sokollik, *Investigations of Field Dynamics in Laser Plasmas with Proton Imaging*, Springer Theses, DOI: 10.1007/978-3-642-15040-1_14,
© Springer-Verlag Berlin Heidelberg 2011

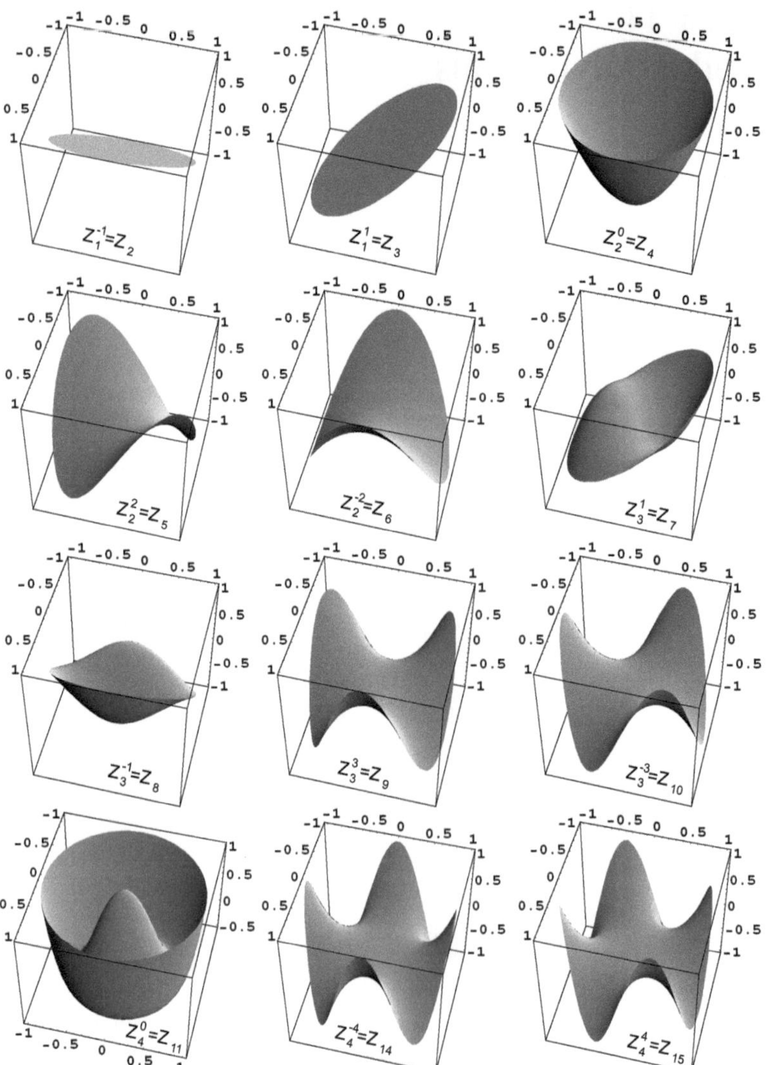

Fig. 14.1 First orders of the Zernike polynomials defined by Eqs. 14.1 and 14.2 calculated with Mathematica

Reference

1. R.J. Noll, Zernike Polynomials and Atmospheric-Turbulence. J. Opt. Soc. Am. **66**(3), 207–211 (1976)

Chapter 15
Gated MCPs

In the imaging experiments presented in this thesis MCPs were used mainly. It was necessary to gate the MCP to avoid influences of signals caused by electrons or X-rays and to achieve an energy selection of the proton signal. Thus, the detector is sensitive in a short time window only (gating time). Due to the broad energy distribution the proton bunch is temporally stretched during the propagation. For the interpretation of the imaging pictures the gating time and the energy of the detected protons signal have to be known. Therefore proton spectra were measured using a Thomson spectrometer and the gated MCPs. In the following the analysis of these spectra will be discussed which delivers the gating time and the proton energy.

Two different gating units were used in the experiments, a commercial gating MCP system (Schulz Scientific Instruments) and a MCP actuated by two high voltage switches for MCP and phosphor screen. The commercial system was used with a fixed gating time. The reference spectra are shown in Fig. 15.5 for different trigger times (time between laser-target interaction and activation of the MCP). The energy resolution of a Thomson spectrometer is defined roughly by the width of the pinhole projection and thus by the vertical profile of the trace. Since the horizontal and vertical width are comparable for the shown spectra the pinhole projection has to be considered to determine the energy resolution (cf. Fig. 15.1a). Thus, the measured signal S is a convolution of the pinhole function P and the spectra E.

$$S = E \otimes P \tag{15.1}$$

The pinhole function is plotted in Fig. 15.1b dependent on the y-coordinate. The fit with a super-gaussian function ($f = \exp[x^4/d^2]$) was convoluted with a gaussian function to reconstruct the measured spectra trace. This gaussian function can be converted into the energy space representing the proton spectra $E(E_{\mathrm{prot}})$. Being aware of the spectra the gating and exposure time of the object to probe can be

T. Sokollik, *Investigations of Field Dynamics in Laser Plasmas with Proton Imaging,* Springer Theses, DOI: 10.1007/978-3-642-15040-1_15,

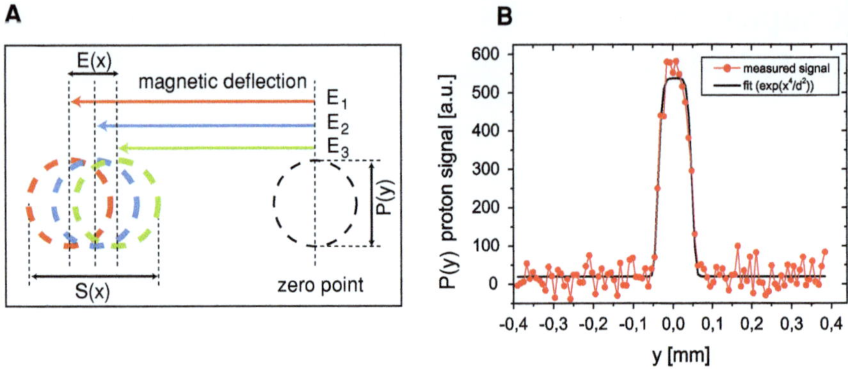

Fig. 15.1 **a** The measured signal $S(x)$ can be described as a convolution of the pinhole function $P(y) \approx P(x)$ and the spectra $E(x)$. **b** Pinhole function $P(y)$ fitted with a super-gaussian function

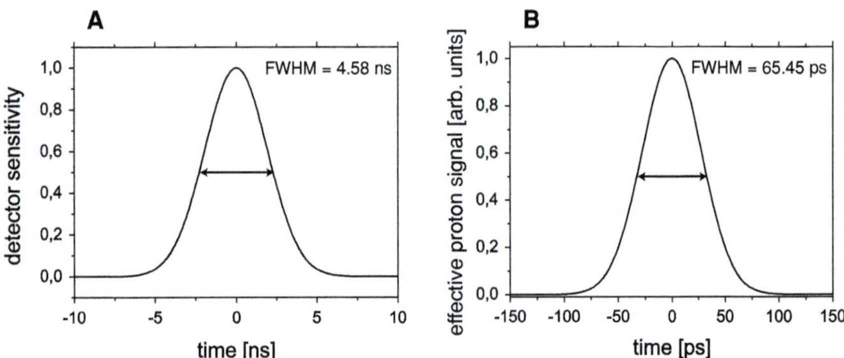

Fig. 15.2 **a** Reconstructed function of the detector sensitivity (gating time). **b** Time of flight ($t = 0$ defined as arrival of the max. proton signal) of the detected proton bunch at $d = 15$ mm

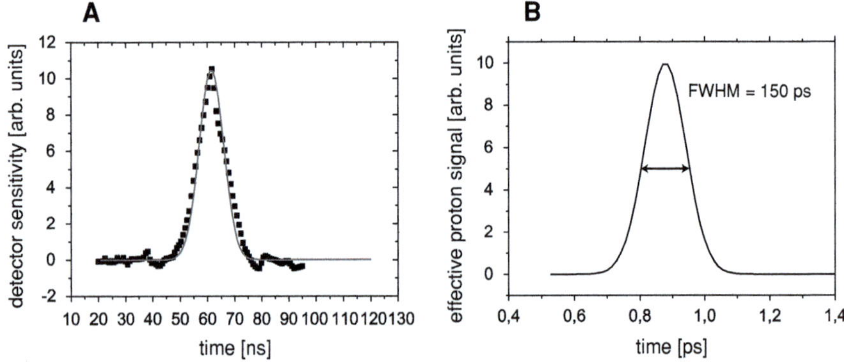

Fig. 15.3 **a** Estimated detector sensitivity fitted with gaussian function. **b** Time of flight of the proton bunch at $d = 15$ mm

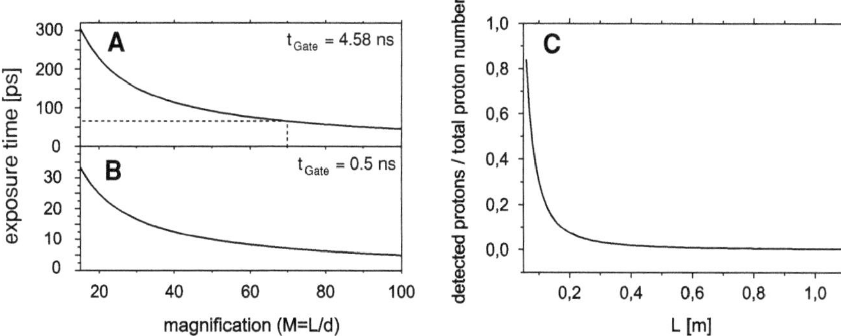

Fig. 15.4 a Exposure time dependent on the magnification for the gating time realized with the commercial MCP system ($t_{Gate} = 4.58$ ns). b Exposure time for a gating time of $t_{Gate} = 0.5$ ns. c Fraction of the detected proton numbers for an MCP with a diameter of 4 mm. (half angle of proton emission 20°) where L is the distance between source and detector

calculated (cf. Fig. 15.2). The gaussian shape of the detector sensitivity can be explained by the high voltage pulse generated by the pulse generator.

In Chap. 9 and in some experiments in Chap. 10 a gated MCP was used with an adjustable gating time. The MCP voltage and the phosphor voltage can be switched on and off independently. Thus, the gating time and the exposure time varies in the experiments. The achieved exposure time was about 400 ps (Chap. 9) and 150 ps (Chap. 10) dependent on the magnification of the images (cf. Fig. 15.3). The exposure time for a fixed time window depends only on the distance of the object and the detector (cf. Sect. 9.3). In Fig. 15.4a the exposure time against the magnifications ($M = L/d$) is plotted for the achieved gating time of $t_{Gate} = 4.58$ ns. To give a prospect for further experiments a graph is shown for a gating time of $t_{Gate} = 0.5$ ns since gating units are available now with a sub-nanosecond gating time. As shorter the gating time as less number of protons can be detected. Thus, the distance to the detector has to be decreased to detect a larger amount of protons and the distance to the object to probe has to be decreased to realize the necessary magnification and exposure time (cf. Fig. 15.4b).

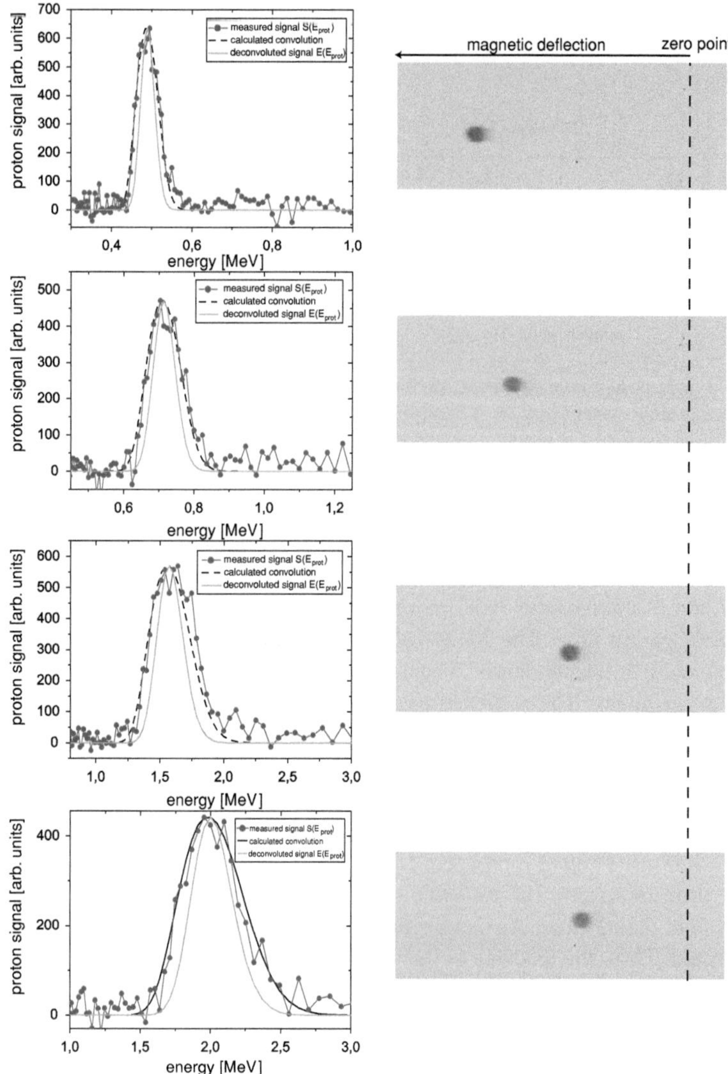

Fig. 15.5 On the *right-hand side* the measured proton signals are shown. The *dashed line* indicates the point of no deflection whereas the deflection of the proton signal due to the magnetic field is indicated by the *arrow*. From the distance of the signals to the zero points the energy can be determined. On the *left-hand side* the corresponding spectra are shown. The measured signal is reconstructed by a convolution of the pinhole function and an assumed spectra. In the time-space these spectra function is constant for this experiment series and defines the detector sensitivity (cf. Fig. 15.2)

Curriculum Vitae

Name: Thomas Sokollik

Place of Birth: Merseburg, Germany

Date of Birth: December, 1979

Current Position

Postdoctoral researcher
LOASIS Program
Accelerator & Fusion Research Division
Lawrence Berkeley National Laboratory
1 Cyclotron Road, Berkeley,
California 94720
Email: TSokollik@lbl.gov

Education and Work History

- Postdoctoral researcher at the University of California Berkeley and Lawrence Berkeley National Laboratory (since 2009)
- Postdoctoral researcher at the Max-Born-Institute for Nonlinear Optics and Short Pulse Spectroscopy (2008-2009)
- Ph. D. in Physics at the Max-Born-Institute for Nonlinear Optics and Short Pulse Spectroscopy and the Technical University of Berlin, advisor Prof. W. Sandner, *Investigations of Field Dynamics in Laser Plasmas with Proton Imaging* (2008)
- Diploma at the Julius-Maximilians University Würzburg, advisor Prof. C. Spielmann, *Generation of High Harmonics using ultra-short, dichromatic laser pulses* (2005)

List of Publications

M. Schnürer, T. Sokollik, S. Steinke, P. V. Nickles, W. Sandner, T. Toncian, M. Amin, O. Willi, and A. A. Andreeva, *Influence of Ambient Plasmas to the Field Dynamics of Laser Driven Mass-Limited Targets.* AIP Conference Proceedings, Vol. **1209**, 111 (2010).

S. Steinke, A. Henig, M. Schnürer, T. Sokollik, P. V. Nickles, D. Jung, D. Kiefer, R. Horlein, J. Schreiber, T. Tajima, X. Q. Yan, M. Hegelich, J. Meyer-Ter-Vehn, W. Sandner, and D. Habs, *Efficient ion acceleration by collective laser-driven electron dynamics with ultra-thin foil targets.* Laser Part. Beams **28**, 215–221 (2010).

A. Henig, S. Steinke, M. Schnürer, T. Sokollik, R. Hörlein, D. Kiefer, D. Jung, J. Schreiber, B. M. Hegelich, X. Q. Yan, J. Meyer-ter-Vehn, T. Tajima, P. V. Nickles, W. Sandner, and D. Habs, *Radiation–Pressure Acceleration of Ion Beams Driven by Circularly Polarized Laser Pulses.* Phys. Rev. Lett. **103**, 245003 (2009).

T. Sokollik, M. Schnürer, S. Steinke, P. V. Nickles, W. Sandner, M. Amin, T. Toncian, O. Willi,*Directional laser driven ion-acceleration from microspheres.* Phys. Rev. Lett. **103**, 135003–135004 (2009).

T. Sokollik, M. Schnürer, S. Ter-Avetisyan, S. Steinke, P. V. Nickles, W. Sandner, M. Amin, T. Toncian, O. Willi, and A. A. Andreev, *Proton Imaging of Laser Irradiated Foils and Mass-Limited Targets.* 2nd International Symposium on Laser-Driven Relativistic Plasmas Applied for Science, Industry, and Medicine, (AIP, Kyoto (Japan)), Vol. **1153**, pp. 364–373 (2009).

P. V. Nickles, M. Schnürer, S. Steinke, T. Sokollik, S. Ter-Avetisyan, A. Andreev, and W. Sandner, *Generation and manipulation of proton beams by ultra-short laser pulses.* 2nd International Symposium on Laser-Driven Relativistic Plasmas Applied for Science, Industry, and Medicine, (AIP, Kyoto (Japan)), Vol.1153, pp. 140–150 (2009).

S. Ter-Avetisyan, M. Schnürer, T. Sokollik, P. V. Nickles, W. Sandner, U. Stein, D. Habs, T. Nakamura, and K. Mima, *Electron sheath dynamics and structure in intense laser driven ion acceleration.* Eur. Phys. J. Special Topics **175**, 117–121 (2009).

P. V. Nickles, M. Schnürer, T. Sokollik, S. Ter-Avetisyan, W. Sandner, M. Index Amin, T. Toncian, O. Willi, and A. Andreev, *Ultrafast laser-driven proton sources and dynamic proton imaging .* J. Opt. Soc. Am. B 25, B155 (2008).

S. Ter-Avetisyan, M. Schnürer, P. V. Nickles, T. Sokollik, E. Risse, M. Kalashnikov, W. Sandner, and G. Priebe, *The Thomson deflectometer: A novel use of the Thomson spectrometer as a transient field and plasma diagnostic.* Rev. Sci. Instruments **79**, 033303 (2008).

S. Ter-Avetisyan, M. Schnürer, T. Sokollik, P. V. Nickles, W. Sandner, H. R. Reiss, J. Stein, D. Habs, T. Nakamura, and K. Mima, *Proton acceleration in the electrostatic sheaths of hot electrons governed by strongly relativistic laser-absorption processes.* Phys. Rev. E **77**, 016403 (2008).

T. Sokollik, M. Schnürer, S. Ter-Avetisyan, P. V. Nickles, E. Risse, M. Kalashnikov, W. Sandner, G. Priebe, M. Amin, T. Toncian, O. Willi, and A. A. Andreev, *Transient electric fields in laser plasmas observed by proton streak deflectometry.* Appl. Phys. Lett. **92**, 091503 (2008).

P. V. Nickles, M. Schnürer, S. Steinke, T. Sokollik, S. Ter-Avetisyan, W. Sandner, T. Nakamura, M. Mima, A. Andreev, *Prospects for ultrafast lasers in ion-radiography.* AIP Conference Proceedings, *submitted*

T. Nakamura, K. Mima, S. Ter-Avetisyan, M. Schnürer, T. Sokollik, P. V. Nickles, and W. Sandner, *Lateral movement of a laser-accelerated proton source on the target's rear surface.* Physical Review E **77**, 036407 (2008).

P. V. Nickles, S. Ter-Avetisyan, M. Schnuerer, T. Sokollik, W. Sandner, J. Schreiber, D. Hilscher, U. Jahnke, A. Andreev, and V. Tikhonchuk, *Review of ultrafast ion acceleration experiments in laser plasma at Max Born Institute.* Laser Part. Beams **25**, 347 (2007).

A. V. Brantov, V. T. Tikhonchuk, O. Klimo, D. V. Romanov, S. Ter-Avetisyan, M. Schnürer, T. Sokollik, and P. V. Nickles, *Quasi-mono-energetic ion acceleration from a homogeneous composite target by an intense laser pulse.* Phys. Plasmas **13**, 10 (2006).

S. Ter-Avetisyan, M. Schnürer, P. V. Nickles, M. Kalashnikov, E. Risse, T. Sokollik, W. Sandner, A. Andreev, and V. Tikhonchuk, *Quasimonoenergetic deuteron bursts produced by ultraintense laser pulses.* Phys. Rev. Lett. **96**, 145006 (2006).

S. Skupin, G. Stibenz, L. Berge, F. Lederer, T. Sokollik, M. Schnürer, N. Zhavoronkov, and G. Steinmeyer, *Self-compression by femtosecond pulse lamentation: Experiments versus numerical simulations.* Phys. Rev. E **74**, 056604 (2006).

Index

A
absorption mechanisms, 25
acceleration
    front-side, 32
    rear-side, 32
Alfvén limit, 30
Amplified Spontaneous Emission (ASE), 37
 autocorrelation, 40

B
bandwidth-limited pulse, 8
Boltzmann distribution, 21
Bragg-peak, 72
Brunel absorption, 27

C
carrier-envelope phase, 8
Ĉerenkov light, 52
charge compensation, 92
charge separation, 22, 25
chirped pulse, 7
Chirped Pulse Amplification (CPA), 9
contamination layer, 30
convolution, 113
CR39 plate, 57
critical density, 17
critical surface, 26

D
Debye length, 20
Debye sphere, 21
deformable mirror, 39
density profile, 26
droplet diameter, 86

droplet generation, 85
droplet spacing, 86

E
electron currents, 51, 30
electron density, 21, 17
electron distribution, 20
electron sheath, 30
electron trajectory, 12
electronic pressure, 20
emittance, 47
energy dependence
          virtual source, 67
exposure time, 74, 115

F
focus distribution, 40

G
gating time, 73, 113
group velocity, 19

H
hole boring, 28

I
ion front, 22
ion spectra, 47

J
$j \times B$ heating, 28

L
laser filaments, 86
laser induced transparency, 18
laser intensity, 10
light pressure, 28
Lorentz equation, 10

M
magnification, 57
mass-limited, 83
mesh projection, 57
monoenergetic deuteron
          spectra, 50
monoenergetic ions, 50
multi-channel plate (MCP), 73
     gating, 73
     time resolution, 74, 113

N
Nd:glass laser system, 40

O
optical probing, 18
overdense plasma, 18

P
parabolic mirrors, 39
particle tracer (GPT), 91
particle tracing, 91, 103
phosphor screen, 73
plasma expansion, 21
plasma frequency, 17
plasma mirror, 18
plasma scale length, 26
plasma shadowgraphy, 43
Poisson equation, 20
ponderomotive acceleration, 28
ponderomotive force, 28
ponderomotive potential, 28, 25
ponderomotive scattering, 12, 19
ponderomotive self-focusing, 19
precursor electrons, 88, 79, 30
proton beam pointing, 51, 103
proton bunch, 74
proton deflectometry, 71
proton divergence, 66
proton images
     2-dimensional (droplets), 89
     2-dimensional (foil), 78
     streak images, 99

proton layer shape, 64
proton spectra, 47, 116
pulse compression, 9

R
radiochromic films stacks, 72
refractive index, 18, 19
relativistic mass increase, 18
relativistic profile steepening, 19
relativistic self-focusing, 19
resonance absorption, 25

S
self-phase modulation (SPM), 9
self-similar solution, 22
Shack-Hartmann sensor, 39
shock acceleration, 32
slowly varying envelope
          approximation, 8
source size, 64, 56
spectral bandwidth, 8
spectral phase, 8
streak camera, 41
streak deflectometry, 97
     proton streak camera, 97
synchronization
          (laser systems), 40

T
Target Normal Sheath
          Acceleration (TNSA), 30
temporal contrast, 26, 33
     Nd:glass laser, 40
     Ti:Sa laser, 37
Thomson spectrometer, 47
     energy resolution, 49
Ti:Sa laser system, 37
time of ight (tof), 44

U
underdense plasma, 17

V
$v \times B$ term, 28, 27
vacuum heating, 27
vector potential, 10
     normalized, 10
virtual source, 57
     virtual source dynamics, 61

W
wave front, 39
Weibel instability, 30, 87

Z
Zernike polynomials, 39, 111